# Introduction: Aim of the Seminar

Over many years international organizations, national organizations, and private concerns have prepared standardized methods of analysis for food products, including milk and milk products, for purposes of quality control, assessment of nutritive content, enforcement of legal requirements, and affirmation of safety. This activity is concerned with identifying the most appropriate current methodology and codifying it in authoritative documents.

The object of the seminar was to appraise the problems that will be faced by analysts of dairy products in the future and examine the means that are likely to be used to solve them. The seminar was thus concerned with the application of methods in quality control and with the techniques that will be in regular use in control, whether highly sophisticated or not and whether automated and/or indirect in principle.

The thoughts of practising analysts in this open-ended forum were expected to stimulate new ideas in the participants. The seminar provided an opportunity for extensive exchange of information and ideas between them.

The programme was based on four themes each comprising a half-day session. Authoritative specialists were invited to present overview papers. Short contributed papers and poster sessions were also presented on each theme. The sessions were complemented by a half-day visit to dairy and food laboratories in which the latest in quality control techniques are employed.

**SCIENTIFIC PROGRAMME COMMITTEE**

*Chairman:* H. Werner, Government Research Institut for Dairy Industry, Hillerod, Denmark, President of IDF Commission E (Analytical standards, laboratory techniques)

G. Steiger, Federal Dairy Research Station, Liebefeld-Bern, Switzerland

F.M. Luquet, Institut Scientifique d'Hygiène Alimentaire, Longjumeau, France

P.B. Czedik-Eysenberg, Oesterreichische Unilever, Vienna, Austria, Chairman of FECS—WPFC

W. Baltes, Food Chemistry Institute, Technical University Berlin, Germany F.R.

G.C. Cheeseman, National Institute for Research in Dairying, Shinfield, U.K.

M. Tuinstra-Lauwaars, Association of Official Analytical Chemists, Bennekom, Netherlands

.R.W. Weik, Food and Drug Administration, Washington D.C., USA

*Secretary:* E. Hopkin, IDF General Secretariat, Brussels.

**ORGANIZING COMMITTEE (United Kingdom)**

The Committee has been formed by the Food Chemistry Group of the Royal Society of Chemistry which, as UK member of FECS, is host for the meeting.

*Chairman:* G.C. Cheeseman, National Institute for Research in Dairying, Shinfield

H. Egan, Association of Official Analytical Chemists, United Kingdom and Ireland

P.J. Frazier, Dalgety Spillers Research Cambridge (Chairman, Food Chemistry Group).

G.R. Fenwick, Food Research Institute, Norwich (Hon. Secretary, Food Chemistry Group)

F. Harding, Milk Marketing Board, Thames Ditton

*Secretary:* E.S. Wellingham, Royal Society of Chemistry, London

Special Publication No. 49

# Challenges to Contemporary Dairy Analytical Techniques

The Proceedings of a Seminar organised by the Food Chemistry Group of the Royal Society of Chemistry in association with the International Dairy Federation, the Federation of European Chemical Societies Working Party of Food Chemistry (FECS Event No. 65), and the Association of Official Analytical Chemists

University of Reading, England, March 28th—30th, 1984

The Royal Society of Chemistry
Burlington House, London W1V 0BN

**British Library Cataloguing in Publication Data**

Challenges to Contemporary Dairy Analytical Techniques.—
   (Special Publication/Royal Society of Chemistry;
   ISSN 0260-6291, No. 49)
   1. Dairy Products — Analysis
   I. Royal Society of Chemistry. *Food Chemistry Group.*     II. Series
   637'.01'543      SF250.5

ISBN 0-85186-925-4

Printed in Great Britain by
Whitstable Litho Ltd., Whitstable, Kent

# Contents

# Control of Systematic Error of Methods of Analysis through Collaborative Studies

By W. Horwitz

FOOD AND DRUG ADMINISTRATION, 200 C STREET, S.W., WASHINGTON, D.C. 20204, U.S.A.

Among the first chapters of most quantitative chemical analysis textbooks is one dealing with analytical error. Some of these texts emphasize the point with shocking examples of how other analysts out in the cruel world of practical analytical chemistry who did not pay attention to their laboratory technique could produce results which did not check with the "correct" values. Not until I joined the Food and Drug Administration a few years after taking quantitative analysis did I discover that obtaining two identical values was the exception, rather than the rule. I also discovered that if you did not want to expose yourself to variability, you should never run more than one analysis per sample.

Contemporary analytical techniques in dairy science, or for that matter in all sciences which utilize chemical analyses, are characterized by a major development which began about 40 years ago. This development is the introduction of instrumentation which permits conducting routine analyses at relatively low concentrations. Today the trend is continuing, not so much into radically new instruments or techniques nor even toward lower concentrations, but extending in ever widening circles into new areas, as for example, automation and computer control.

I have learned from sad experience never to be a prophet; as soon as you make a prediction, nature has a way of taking off in an entirely unexpected direction. Yet as overview speakers, my associates and I are expected to convey words of wisdom which should make your attendance at this symposium worth while. I have therefore chosen for my overview topic to review some practical and administrative aspects of modern analytical chemistry which might lead to improvements in the practice of analytical chemistry through the conduct of collaborative studies.

The job of a scientist is to produce useful information in its broadest sense. In the analytical area, the information sought deals with the chemical or biological composition of materials. As the production of dairy products and other foods shifted from the producer being his own major consumer to producing for sale to anonymous consumers, it became the function of government to assure citizens of the safety of products produced by others as well as to guarantee their economic value in areas where the consumer could not protect himself. When the manufacturer was made responsible for keeping the composition of his products within the limits set by governmental regulation or economic contracts, analytical chemists became the arbiters and enforcers of the standards and tolerance requirements.

In this role of enforcement of specifications, analytical chemists found that although each one could check himself nicely, other chemists might not check his results. Obviously, it was the other chemist who was wrong and a third chemist was often brought in as an arbiter. In many cases, the new laboratory merely provided a different answer. Chemists generally disassociated themselves from being the cause of the disparate results and assigned the cause to external forces, such as poor sampling practices on the part of the sample collectors or the fact that each laboratory used a different method of analysis.

However, most of us are more familiar with methods of analysis than with methods of sampling. A sample is delivered to the laboratory and the report reflects the composition of that laboratory sample, or, more appropriately, the composition of the test portion removed from the test sample prepared from the laboratory sample. But we have a long way to go in improving methods of analysis without venturing into the even more variable subject of sampling.

Therefore, I will discuss specifically the search for "improvements" in methods of analysis and how we dissipate our efforts in searching for minor improvements because we do not know how to extract the clues that lead to major improvements. I will take as an example a very important method of analysis, the determination of phosphorus by the formation and reduction of phosphomolybdate to the familiar molybdenum blue, whose absorbance is measured colorimetrically. Phosphorus is one of the backbones of the fertilizer industry and is an element fundamental to the internal energy relationships of living organisms. The determination of phosphorus, therefore, is not a trivial exercise.

## An Example of Systematic Error - Phosphorus in Cheese

As many of you know, for about 20 years there has existed a cooperative effort on the part of the International Dairy Federation (IDF), the International Organization for Standardization (ISO), and the Association of Official Analytical Chemists (AOAC) to supply the Codex Alimentarius Commission of the Food and Agriculture Organization (FAO)/World Health Organization (WHO) of the United Nations with the methods of analysis required for their program of establishing standards for milk and milk products. As part of this effort, the AOAC conducted a collaborative study[1] which included the determination of phosphorus in cheese by the ISO molybdenum blue method[2].

First Collaborative Study. The study involved 5 levels in duplicate of phosphorus in cheese products, 2 at about 0.4% and 3 at about 0.8%, examined by 7 laboratories. The individual values for each set of duplicates from each cheese (products 1-5) from each laboratory (labeled 1-7) and the precision measures are given in Figure 1, arranged in ascending order of laboratory averages. The precision measures plotted are the ISO repeatability (r) and reproducibility (R) intervals, given in the same absolute values of percent phosphorus on the left vertical axis labeled "%P." The corresponding coefficients of variation, a relative measure, are read against the right vertical axis labeled "CV, %". With a little imagination, you can see that the indices based on the standard deviations (r, R) increase a little with increasing concentration, whereas the coefficients of variation (CV) are decreasing with increasing concentration. There were no individual value or laboratory average outliers in the series of 70 values, but 2 of the laboratories were rank sum outliers. This means that the consistently highest (laboratory 2) or lowest (laboratory 5) position over all 5 products would have occurred by chance alone only once in 20 or more times. These laboratories

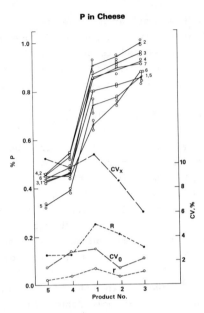

Figure 1. The data (duplicates) from the first collaborative study of the determination of phosphorus in cheese products[1] by the colorimetric molybdenum blue method ISO 2962 and the derived precision indices (ISO repeatability interval, $\underline{r}$, ISO reproducibility interval, $\underline{R}$, read in %P on the left scale, and the corresponding coefficients of variation, CVo and CVx, read on the right scale). The numbers adjacent to the lines connecting the data are the laboratory numbers.

---

were not removed, since to have done so would involve loss of almost 30% of the data, which is considered excessive. The conclusion from this study was that the repeatability was good but that the method required further work to reduce the large among-laboratories variability.

Second Collaborative Study. A second study was conducted a few years later by Dr. Steiger of Switzerland under the auspices of IDF[3]. The ISO method was slightly modified to use ascorbic acid instead of hydrazine as the reducing agent and the wavelength of measurement of the color was shifted from 700 to 820 nm. Ten

laboratories examined 12 products all at approximately 0.9% P, by the original method and by the modified method. Since there was no statistically significant difference between the two methods, the data were pooled to provide 4 replicates from each laboratory for each product. The average of the replicates was plotted in Figure 2 together with $CV_x$ and R. Since all of the levels examined contain approximately the same phosphorus content, the precision measures are approximately constant. Again 2 laboratories (numbers 3 and 11) are outliers by the rank sum test, but this time with a probability of less than 1 in 1000 that either of them would consistently rank lowest and highest, respectively, among 10 laboratories by chance alone in a series of 12 products. Although Dixon and Cochran outliers were checked for and found in the data, usually in data from laboratories 3 and 11, these outlier tests are not particularly pertinent in this case. The Dixon test does not appear to be appropriate since whether or not a value is an outlier by this test depends heavily on the position of its nearest neighbor and the next to the last opposite neighbor. For each material, 8 values are more or less centrally bunched, and not randomly distributed. The Cochran test also does not appear to be particularly appropriate because most of the within- laboratory variability indices are very small so that even a mild deviation becomes "significant." The occasional apparent aberration, such as the CVo for product 2 and the CVx for product 7, are to be expected every so often due to random fuctuations. The CVx of about 7% is somewhat better than the 9% of the previous study but the difference is not statistically significant. In any case, a reproducibility (ISO) of 0.2% (absolute) between-
laboratories at the 1% phosphorus level will not permit much regulatory control of added phosphate to cheese products. A product would have to contain more than an apparent 1.2% P before a regulatory official would consider the product in violation at the 95% confidence level. He would have to allow another 0.1% if he wished to use the 99% confidence level.

Interpretation. These limitations on regulatory action are not the most interesting aspect of the second study; the most important conclusion is that the second study verified the susceptibility of this method to laboratory-induced variations.

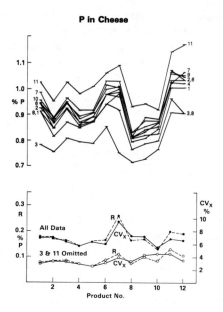

Figure 2. The pooled data (laboratory averages of 4 determinations) from the second collaborative study[3] of the determination of phosphorus in cheese products by the original and slightly modified colorimetric molybdenum blue method ISO 2962 and the reproducibility interval, R, and reproducibility coefficient of variation, CVx, with and without the rank sum outlying laboratories 3 and 11 omitted. The individual data points and the within-laboratory variability precision indices have been omitted for clarity. The CVx and laboratory numbers are read as in Figure 1.

Although results from most of the laboratories were fairly well bunched within a spread of about 0.1%, the values from the 2 extreme laboratories expanded this spread to about double this value. Two independent studies now have shown that laboratories introduce systematic errors into their analyses without being aware that there is anything wrong. Furthermore, it is to be noted that there was also a tendency in both studies for a laboratory to maintain its position relative to the other laboratories consistently throughout the entire series. The most likely cause of this type of consistent error, across as many as 12 materials analyzed in a series, lies in the preparation of the standard solution and in the construction of the calibration curve.

It is now too late to request the collaborating laboratories to examine their standard solutions for possible deviations. I strongly suspect that, particularly with the highest and lowest laboratories, we would find that the discrepancies between their standard solutions would account for most of the differences. A similar conclusion was reached as a result of a review of the data from the Smalley proficiency studies[4] of aflatoxin in seed crops sponsored by the American Oil Chemists' Society. On two successive years, a solution of pure aflatoxin, which merely required spotting, developing, and comparison with the standards prepared by the individual analysts, showed a higher among-laboratories variability than the test materials themselves. In another instance, Brown and his colleagues[5] found it necessary to require the preparation of 5 separate independent calibrating solutions from National Bureau of Standards (NBS) Certified Calcium Carbonate in their final reference method for the determination of calcium in blood serum. Separate weighings for each standard solution were required to permit some cancellation of the random errors involved in the preparation of calibrating solutions.

## An Example of the Reduction of Systematic Error

One of the most conclusive pieces of evidence that chemists do not know how to prepare accurate standard solutions is provided by the collaborative study of the determination of extractable color in capsicums and oleoresin[6]. Originally, the reference standard was a colored solution prepared from specified weights of potassium dichromate and cobalt ammonium sulfate with an absorbance in a 1 cm cell at 460 nm of 0.600. This solution was used to calibrate the spectrophotometer in view of the inability to prepare a suitable stable, natural-color standard. Later, a colored glass with an absorbance certified by NBS was permitted as an alternative. The use of the two standards was compared on two materials in a collaborative study, with the results shown in Table 1. The within-laboratories coefficients of variation using

Table 1. Results from the collaborative study of the determination of color units of extractable color in oleoresin paprika

|  | Sample 1 | | Sample 2 | |
|---|---|---|---|---|
|  | Glass Std | Chemical Std | Glass Std | Chemical Std |
| Mean, arbitrary units | 996 | 1027 | 2012 | 2078 |
| Repeatability coefficient of variation, % | 0.9 | 0.9 | 0.5 | 0.6 |
| Reproducibility coefficient of variation, % | 1.4 | 4.8 | 2.2 | 7.8 |

either standard were about the same, 0.7%, for both materials. The reproducibility coefficient of variation, however, was reduced by a factor of about 3 by substitution of the certified glass filter for the laboratory-prepared chemical-based reference solution.

In the light of these studies, I suggest that the solution to the persistent systematic error of the determination of phosphorus in cheese lies in the reduction of the systematic error. This reduction can be accomplished by incorporating a pertinent physical constant into the method as a systematic error quality control point. The control could take the form of a statement in the method that the reference standard solution of phosphorus, diluted to an indicated concentration, when conducted through the method should give an absorbance in a 1 cm cell of x, plus or minus a reasonable tolerance. The value of x and its tolerance can be obtained as part of the collaborative study by requesting the collaborators to report the absolute absorbance value obtained from a specified working standard solution. Alternatively this value can be obtained directly by requesting submission of copies of the collaborators' standard calibration curves and noting the concentration corresponding to a convenient absorbance value in the region of 0.4-0.6, where the spectrophotometric measurement can be made with greatest reliability. Various refinements can also be made such as requiring calibration of spectrophotometers with glasses certified for absorbance, or noting the slope and intercept of the calibration curve. Obviously in measurements taken with recorders and associated electronics which permit baseline shifts, scale expansion, and other manipulations by changing the settings, reference standards must be run with the same instrument settings used for the assays and with sufficient frequency to detect drift in instrument response. Our instrument chemists are clever enough to develop the appropriate quality control procedures once they understand the nature and importance of the problem.

For many years, analytical chemists have been concentrating on reducing the within-laboratory component of analytical error, which is usually the less important part of analytical variability. Within-laboratory variability (repeatability) is also the most accessible part of the total variability since it is entirely within the control of the individual analyst. Control of among-laboratories variability requires the utilization of an elaborate organizational structure such as those maintained by the International Organization for Standardization, the Association of Official Analytical Chemists, and the American Society for Testing and Materials. However, the incorporation of a reference physical constant does not require the elaborate logistics of a collaborative study. In the case of the phosphorus example, a simple request to a dozen laboratories, for example, to determine the absorbance of an x mg/L solution of dipotassium hydrogen phosphate by the specified method would probably provide a response on the part of most of them. An acceptable specification interval could be obtained by utilizing a censored mean[7] (removing values more than 3 standard deviations from the average) or a trimmed range[8] (routinely removing the highest and lowest values), and rounding to convenient values. Parenthetically, it should be kept in mind that molybdenum blue is not a simple compound. The color intensity is sensitive to small changes in operating conditions and this may complicate the separation of random and systematic error in this particular case.

## Comparison of Methods

Many laboratories are spending too much time on marginal improvements in methods of analysis and in contests to determine which of several methods is superior. Laboratory administrators have not yet discovered the best kept secret of statistics: estimates of precision are grossly imprecise. The magnitude of this uncertainty can be determined, depending on the design of the study, i.e., the number of levels, laboratories, and replicates involved. Most ISO methods call for the performance of duplicates, so this degree of replication will be considered. Improvement with greater replication is marginal. Table 2 gives the percentage factors of the square root of the variance components (equivalent to the standard deviation of the standard deviation) for the within- ($S_o$) and among-laboratories (not including the within-) ($S_L$) components of error as reported to the AOAC's Committee on Interlaboratory Collaborative Studies.

Table 2. Approximate relative (as percent) standard deviation of $S_o$ and $S_L$ as a function of the number of laboratories contributing duplicate determinations

| No. of Labs | Within-laboratories $S_{o,}$ % | Among-laboratories $S_{L,}$ % |
|---|---|---|
| 2 | ±84 | ±113 |
| 3 | 79 | 105 |
| 5 | 73 | 95 |
| 7 | 69 | 89 |
| 10 | 64 | 82 |
| 20 | 55 | 70 |
| 40 | 47 | 59 |
| 100 | 37 | 47 |

A statistician may disown this table on the grounds that his theories apply only to variance, not to the standard deviation. but his reservations are of little practical concern in view of the point being made--that the variability is huge. For a 6 laboratory collaborative study, with each laboratory performing duplicates, the relative standard deviation of the estimate of the repeatability is about 70%; the standard deviation of the among laboratories standard deviation (not including within-laboratories) is about 90%. Even accepting the crudeness of these estimates, they indicate that standard deviations must differ by a factor of the order of magnitude of about 2 before we can even consider the existence of a significant difference in these precision measures. If we now consider how many times we have made selections of methods based upon differences in standard deviations of the order of 10%, it is easy to see why controversies between methods are rarely settled by a few experiments in a single laboratory. In view of this factor, the choice of method in an actual situation is usually a matter of practicality--utilize the cheapest and fastest method with the required degree of reliability.

## Equivalency of Methods

Where the total error (systematic and random) is relatively large as in trace analysis of materials such as pesticide residues, toxic elements, and mycotoxins, the literature is supersaturated with minor modifications of a few basic procedures. As a result of collaborative studies performed with different methods on pesticide residues[9] and aflatoxins[10], we can now see that their performances are more or less equivalent. The choice of method by a laboratory is now becoming more a matter of familiarity, which leads to superior performance with a method that has been in continuous use as compared to a method being tested for the first time in the same laboratory. This is well demonstrated in a recent comparative study of the determination of dioxin in fish[11]. Eight laboratories (one of which had to be rejected as an outlier) achieved a reproducibility coefficient of variation of about 20% for dioxin in fish at the approximately 50 ppt (pg/g) level, each using its own extraction and cleanup procedure, mass spectrometry instrumentation, and isotope-labeled internal or external standard. Although each method was prone to systematic errors in each laboratory (recovery of the internal standard ranged from 29 to 109%), normalization against a supplied standard cancelled a large part of the systematic error. All of the participating laboratories had many years of experience analyzing for dioxins by their local procedure. It could be anticipated that using an unfamiliar method undoubtedly would have led to substantial variability in comparison to the "house" method. It may be concluded that where a large investment in training and experience has been made in methods with a high degree of variability (e.g., reproducibility coefficient of variation in excess of 10%), the local method has a built-in advantage of familiarity. But when a laboratory is starting an analytical program in any unfamiliar area, good management as well as legal counsel usually advise the use of the method employed by the regulatory agency.

These examples of different methods that can lead to the same analytical results can be applied only to situations where expertise has been developed through years of experience. The institutional method has been optimized and has achieved a high degree of ruggedness through empirical correction of deliberate and unexpected abuse. Such a tolerance for misuse cannot be expected of methods not previously used or in the course of development. Efficient management should require sufficient practice and experience with methods already optimized and utilized in other laboratories before concluding that the reported method is unworkable and requires improvement or replacement, an expensive decision at the very least.

## Summary

One of the chief sources of variability in chemical analysis is the systematic error introduced by individual laboratories. This bias is invisible to a solitary laboratory and becomes apparent only in a multi-laboratory environment or through the use of reference standards. Control of this error in routine analysis requires the introduction into methods of physical constants of standard materials as systematic error quality control points. The other major source of error is random error. Precision differences must be on the order of a factor of 2 to be significant, with experiments containing only about a dozen data points. Therefore research on better methods which lead only to marginal improvements is a misuse of resources. In trace analysis where total variability (systematic and random errors) is large, different methods used in different laboratories often lead to equivalent results, particularly when results are normalized through the frequent use of reference standards. This phenomenon is ascribed to a high level of expertise with the institutional method and a lack of familiarity with any newly introduced method to which it is being compared.

## References

[1] H.R. Brzenk and O.J. Krett, J. Assoc. Off. Anal. Chem., 1976, 59, 1142-1145.

[2] International Organization for Standardization, "Cheese and processed cheese products - Determination of phosphorus content (Reference method) ISO-2962(E)", Geneva, Switzerland.

[3] G. Steiger and W. Horwitz, submitted to J. Assoc. Off. Anal. Chem.

[4] J.D. McKinney, J. Assoc. Off. Anal. Chem., 1981, 64, 939-949.

[5] S.S. Brown; M.J.R. Healy; and M. Kearns, J. Clin. Chem. Clin. Biochem., 1981, 19, 395-412; 413-426.

[6] J.E. Woodbury, J. Assoc. Off. Anal. Chem., 1977, 60, 1.

[7] C.H. Perrin and E.M. Glocker, J. Am. Oil Chem. Soc., 1968, 45, 596A.

[8] P.J. Huber, Ann. Math. Statist., 1972, 43, 1041.

[9] J.A. Burke, The Interlaboratory Study in Pesticide Residue Analysis. In H. Frehse and H. Geissbuhler, Eds. Pesticide Residues (Zurich 1978), Perma Press, Oxford, England, 1979, 19-28.

[10] P. Schuller; W. Horwitz; and L. Stoloff, J. Assoc. Off. Anal. Chem., 1976, 59, 1315-1343.

[11] J.J. Ryan; J.C. Pilon; H.B.S. Conacher; and D. Firestone, J. Assoc. Off. Anal. Chem. 1983, 66, 701-707.

# Experiences in the Certification of Major and Trace Elements in Milk Powder

By B. Griepink

COMMISSION OF THE EUROPEAN COMMUNITIES, RUE DE LA LOI 200, B-1049
BRUSSELS, BELGIUM

## 1. ANALYTICAL QUALITY CONTROL

Measurements form the foundation of all public decisions. If
we want our decision to be the best possible we must assure
that their foundation, the measurements, are correct.

Analyses are carried out in trade. Prices of goods or even
the acceptance of goods by the buyer depend on analytical
results. Long term effects on public health or
epidemiological conclusions are based on the results of
analyses carried out over many years in many countries.

Analytical laboratories therefore are concerned to produce
results which can be confirmed by other laboratories. They
spend a large effort in intercomparisons.

But intercomparability does not give any certainty that the
results are accurate. In many cases, the results must be
accurate, e.g. in calculating the total load of nitrate in
public health results of analyses of cheese are combined with
those of vegetables etc.

So analytical laboratories are concerned to use and develop
methods which yield intercomparable results which approach a
"trace value" as closely as possible.

But even when laboratories use very elaborate and accurate
methods the results can be wrong. No result of an analysis
can be better than the technician and the laboratory who made
the result.

For these and many other reasons the control of analytical
quality is a must in a modern laboratory. This control
involves that the best method is critically selected,
critically approved, critically examined, critically applied
and that all possible tools to assure analytical quality are
used frequently.

2.   METHODS OF ANALYTICAL QUALITY CONTROL

The use of an accepted method, e.g. developed in the framework of the International Dairy Federation (IDF), the
Association of Official Analytical Chemists (AOAC) or the
International Standardisation Organisation (ISO), does not
imply automatically that the results are comparable to others
obtained elsewhere with the same method, nor that the results
are accurate.

This is illustrated with figure 1, which represents the
results of a restricted analytical intercomparison of the
content of spiked nitrate in a milkpowder (BCR 150). In fig. 1
the several obtained values are indicated by numbers.  These
numbers stand for reduction over (copperised) cadmium and
photometry (1, 3, 4, 6, 9, 10), gas liquid chromatography of
nitroxylenol (2), spectrometry of nitroxylenol (5, 11, 13),
ion chromatography (8) and isotope dilution mass spectrometry
(12).  The official IDF-methods produced by well-experienced
IDF-member laboratories scatter more than other techniques
like isotope dilution mass spectrometry.

There are various methods and levels at which control of
analytical quality should take place.  The methods briefly
outlined under 2.1 to 2.4 are complementary.  All the
presented methods should be applied.

The most frequent means of control of analytical quality are:

i)    inner-laboratory control; laboratory standard;
ii)   intercomparison;
iii)  standardised methods;
iv)   reference materials.

Figure 1

2.1   Control in the Laboratory

The first level of control is the continuous critical attitude
of the technician and laboratory head.

A careful investigation of all the sources of error in a whole
chain of analyses and a continuous checking of the method's
performance is necessary to obtain accurate results.  Table 1
presents an example of a general check list.  The table does
not intend to be exhaustive in mentioning all details.

TABLE 1  –  Examples of sources of error in trace analysis

| Analytical Step | Possible Errors | Remarks |
|---|---|---|
| Sample preparation | Drying : measurement of dry mass<br>Crushing, grinding, mixing<br>Weighing | In all stages, losses and contamination possible |
| Sample (pre)treatment | Purification of reagents<br>Digestion, ashing, uniform speciation and oxidation state<br>Dilution, concentration procedure, filtration, percolation, extraction etc.<br>Preparation for end-measurement, neutralisation etc. | In all steps, losses and contamination possible |
| End-measurement | Reagents for calibration, (stoichiometry, stability)<br>Calibration<br>Background correction<br>Interferences in the measurement<br>Signal measurement<br>Relation signal to content | In most steps, losses and contamination possible |

Besides all the checks one should apply (e.g. calibration
graph, linearity, application of standard additions in
relation to calibration graph, changing concentrations of
interfering elements, completeness of digestion or clean-up
reproducibility, blanks, etc.), the use of a laboratory
standard should be recommended. This standard is a material
of which the homogeneity and stability are considered to be
adequate. The standard should be analysed frequently in the
course of the normal routine determinations.

Abrupt changes in the obtained results or a systematic drift
of the results give indications on the performance of the
analytical procedure. The use of such a standard only gives
an indication whether the output of the analytical procedure
is constant in time; it does not guarantee that the thus
obtained results are accurate. The reliability increases when
the laboratory standard has been analysed preferably by
another laboratory and also preferably by another procedure
and if the results of both agree.

## 2.2   Intercomparison

Intercomparisons are held for various reasons. In designing
an intercomparison it is important to consider the aim of the
intercomparison, e.g. the performance of a method in various
laboratories, the control of intercomparability of results, an
approximation of a "true value", the selection of laboratories
suited for a certain analytical task, the assessment of the
state of the art of analyses, etc. The design of the
intercomparison will be different for any of these different
aims.

In this paper we consider mainly the control of analytical
quality as a means to improve accuracy of analyses. In this
context, intercomparisons are regarded as means to approximate
the "true value".

Results of one well-defined method obtained from different
laboratories can give an indication of a laboratory's
performance as compared to the performance of colleagues.
There is, however, no certainty at all that the value upon
which all the laboratories agree is a good value or even the
best approximation of the truth.

Every method and every laboratory is likely to make some
systematic errors, which are small in the case of a good
method followed in a good laboratory. Laboratory agreement in
an intercomparison as described above indicates only that the
laboratories were of equal quality in applying the method. It
does not indicate that the method is free of systematic
errors.

Aiming at accuracy one should apply different methods. These
methods must differ in principles and not just in small
variations. If the results of different methods (which all
have different sources of error and procduce different
systematic errors) are in agreement this makes the outcome a

better approximation of the "true value". For example, if the
results of principally widely differing techniques, like NAA
(neutron activation analysis), ICP (inductively coupled plasma
emission spectrometry), AAS (atomic absorption spectrometry)
and spectrometry, agree on the content of iron in milk, one
may conclude that this final result is free of measurable
biases.

Besides intercomparisons to check a laboratory bias (all
laboratories apply a standard analytical procedure),
intercomparisons in order to find an accurate value are
necessary.

Such intercomparisons often are necessary to develop standard
methods.  In such an intercomparison, a draft standard ideally
should be applied besides other techniques to detect a
possible bias in the draft standard.

## 2.3   Standardised methods

Routine methods have the advantage that they suit economically
best in the particular circumstances of a laboratory.  To be
sure about the quality of the results of such a routine method
in the given context, one can participate in an
intercomparison and compare the results with others.  This,
however, is not always practical:  intercomparisons are not
always organised when one needs a check of the method and
usually the evaluation of an intercomparison arrives on the
participant's desk many months - if not more - later.  So, it
actually gives an impression of the laboratory's performance
at a moment which usually is half a year or more ago.

The necessary instantaneous check can be provided by a
comparison of the routine result with the result of a standard
method, both applied on the same sample, preferably on the
same day.

A standard method is a (sometimes more elaborate) method which
is well-studied and which is not sensitive to small
fluctuations in the procedure as carried out in practice.
Also a standard method requires a procedure which can be
followed in principle by all laboratories in a certain type of
analysis.

But there are drawbacks.  No method is better than the
technicians applying it.  If the reference method is used
occasionally to verify the results of the routine procedures,
there is the cost of starting up and technician training.  If
the technician is not experienced enough, the results of the
reference method will be less trustworthy than the results of
the routine method.

The development of a reference method is quite a different and
time consuming task.  Laboratories are not always prepared to
carry out this task.

A standard method, once developed, reflects the best of the art of the moment. Later developments, improvements or new and better techniques are very difficult to incorporate.

Therefore, in conclusion, the availability and application of a standard method is an important contribution to the analytical reliability, but it does not assure analytical reliability. A too strong belief in the method has often caused undetected errors.

## 2.4  Reference materials

As already indicated in part 2.2, the best way to achieve an accurate result is to apply more and independent techniques to analyse a given sample. This is exactly what is done to obtain a value in a reference material suitable for certification. But before one can think of analysis and certification a few demands have to be met by the material in order to allow possible certification.

Without going into details a few demands will be mentioned:

i)  A candidate reference material should be stable in order to be applicable in many laboratories over a longer period. Stability can be checked by repeated analyses. It is sometimes possible to carry out special treatments, the outcome of which have a certain predictive value.

ii) A candidate reference material should be homogeneous. Homogeneity of any sample, and thus certainly of a reference material, depends on the size of the sub-sample taken for analysis. So the homogeneity should be known in dependence of the sub-sample size. The smaller this size the greater the risk of detectable inhomogeneities. Homogeneity should be proven at least at the size of the intake in common analyses.

If these requirements are met certification can be undertaken. Certification involves careful analysis of the material without introducing any detectable systematic error. The avoidance of systematic errors involves that the methodological bias is checked and found to be negligible and that the same is done to avoid any laboratory bias. This means in practice that more independent techniques are used (see 2.2) by more independent laboratories or departments in one laboratory.

Furthermore these laboratories must work under certain conditions so that their findings are traceable to SI-units. This means e.g. that quite an effort is made to ascertain the purity and stoichiometry of used chemicals and calibrands.

If these requirements all are fulfilled and if laboratories and techniques agree on their measured values, we may say that these values are, according to the state of the art, bias-free, which means that they are close approximation of the "true value" (e.g. K. Heydorn 1980). These values then can be called "certified values". (ISO/REMCO 59)

But before becoming really certified values the pertinent
methods have to be examined closely:  is there any
methodological bias likely or detectable; are the applied
methods the most suitable available at a time; are results of
important methods not included; are the results obtained
really the best to be expected from a certain method for this
particular matrix; _etc._   The certifying body can only give a
certified value if a critical examination of all the steps
involved does not reveal doubts to the values.

But then such a certified reference material can be used to
verify any analytical procedure be it a routine procedure or a
specially designed procedure.  In case a reference material
can be applied it presents an efficient and reliable way of
controlling the analytical performance.  However, the
application is restricted to similar matrices as the matrix of
the reference material.  The judgement about the results of a
method loses power when the matrices differ more and more.

Analysing a certified reference material actually means
comparing one's own results with those of the certifying body
or laboratories.

This is similar to participation in special intercomparisons.
But unlike participation in these intercomparisons, the
information on the own performance is immediately received.
Therefore, the availability of suited reference materials
reduces the necessity of participation in many
intercomparisons, thus in turn reducing the high cost of
intercomparison studies.

Given the useful application of certified materials as well as
the importance of the conclusions based on analysing these
materials, it is clear that manufacture and certification
should be carried out by specialised institutes or
organisations with utmost care.

In the dairy industry, certified reference materials can
especially be useful in:

i)     removing hindrances to trade or trade barriers;
ii)    analysis of milk products for the application of laws or
       directives;
iii)   food studies to investigate the total uptake of a certain
       compund or element by a population;
iv)    the development and evaluation of standard methods of
       analysis;
v)     the evaluation of new methods of analysis, _etc._


3.   HOMOGENEITY INVESTIGATIONS IN THE MILKPOWDER SAMPLES

As indicated before the homogeneity of a sample of which the
contents will become certified is crucial.  In agreement wih
the statement made before one should always indicate at which
level (_i.e._ intake) the homogeneity is studied.

The homogeneity of a milk powder sample should be proven for
the matrix components. If the material would consist of
different phases at the level of intake this could be found by
repeated determinations of matrix components using a precise
method.

In case trace elements will be certified, an additional study
should be made on the distribution of some trace elements.
Not all trace elements can be studied because of their
abundance and low concentration. In addition one should use a
method which is precise enough to detect possible
inhomogeneities at a level which is lower than the value
finally to be certified.

The example of the homogeneity studies on BCR 151 (skim milk
powder spiked with heavy metals) will be taken to illustrate
the homogeneity studies necessarily undertaken and their
outcome.

a) Homogeneity studies for matrix components

Calcium was determined in the most reproducible way ten times
in one bottle and then once in twenty different randomly taken
bottles of the material. The intake was about 250 mg.

The repeated measurements on the homogenised content of one
bottle gave a within bottle coefficient of variation which
contains the overall variation of the analytical measurement
and the inhomogeneity in the content of one bottle. The
twenty replicate measurements on twenty bottles give an
overall coefficient of variation which is composed of the
coefficient of variation within bottle and the coefficient of
variation caused by differences between bottles. This last
coefficient of variation gives an impression of the
contribution of a possible inhomogeneity to the analytical
uncertainty.

Table 2 presents some values to obtain an impression of the
inhomogeneity of some of the matrix elements in BCR 151.

TABLE 2

| Element | Intake level (mg) | Within bottle CV(%) found | Overall betw.bottles CV (%) found |
|---------|-------------------|---------------------------|-----------------------------------|
| Ca      | 250               | 0.36                      | 0.65                              |
| N       | 6                 | 0.34                      | 0.43                              |
| P       | 1                 | 0.12                      | 0.65                              |
| Cl      | 1500              | 0.032                     | 0.037                             |

Table 2 shows that inhomogeneity does not depend only on the
level of intake but as well on the particular elements.
Anyhow, the possible inhomogeneities are so low that they will
not affect the uncertainties in the analysis of trace
elements. These uncertainties are in the order of magnitude
of 10% or more. But even if one would like to use a material

with such a homogeneity for the determination of some matrix
components it would be useful because the usually required
repeatibilities according to IDF standards are larger (e.g.
Ca: 7%, P: 3%, Cl: 4%, N: 0.8%).  The contribution of
inhomogeneity to the total analytical spread of one technician
at the same day with the same reagents and instruments is very
small.

So, from its matrix composition the material can be regarded
as being sufficiently homogeneous.  But on this material 151
which is intended for trace metal analysis another check of
the homogeneity at least for some of the trace elements should
be made.

In this example, the test was made applying instrumental
neutron activation analysis.  This technique can be applied in
a reproducible manner, thus allowing to obtain a good
impression of the maximal inhomogeneity for many elements in
one go.

In principle, the technique is interpreted in the same way as
the tests of the matrix components.  Here it is possible to
establish a methodological coefficient of variation from
weighing error and counting statistics.  The overall
coefficient variation contains this methodological coefficient
of variation.  The remaining is a conservative estimation of a
possible inhomogeneity since not all other sources of random
errors are known.  These random errors then contribute to the
obtained inhomogeneity.

Some findings of homogeneity testing are summarised in
table 3.  The concentrations of most of the trace elements
were too small for a measurement.

TABLE 3

| Element | Intake level (mg) | Calc. CV % | Found CV % | Pos. contrib. of inhomog. CV (%) |
|---|---|---|---|---|
| Br | 10 | 2 | 3.4 | 2.7 |
|  | 50 | 3 | 2.4 | < 1 |
|  | 100 | 2.5 | 2.3 | < 1 |
| Cu | 10 | 4 | 8.3 | 7.3 |
|  | 50 | 4 | 6.1 | 4.6 |
|  | 100 | 2.5 | 2.3 | < 1 |
| Fe | 10 | 3 | 5.6 | 4.7 |
|  | 50 | 3 | 4.5 | 3.3 |
|  | 100 | 2 | 1.7 | < 1 |
| Na | 10 | 2 | 1.8 | < 1 |
|  | 50 | 2 | 1 | < 1 |
|  | 100 | 2 | 1.7 | < 1 |
| Rb | 10 | 3.5 | 5.9 | 4.7 |
|  | 50 | 2 | 2.0 | < 1 |
|  | 100 | 2 | 2.2 | < 1 |
| Zn | 10 | 3 | 5.5 | 4.6 |
|  | 50 | 2.5 | 3.2 | 2.0 |
|  | 100 | 2 | 2.2 | < 1 |

Table 3 shows clearly that the contribution of a possible
inhomogeneity to the total uncertainty depends on the level of
the sample intake. The few elements tested show further on a
possible inhomogeneity which is far within the normal errors
of analysis (the IDF standard for Na quotes a repeatability of
7%, the errors in trace analyis unsually are higher than 10%)
even at such levels of intake of 50-100 mg.

BCR nr 151 has been chosen as an example to illustrate the
care which should be taken for the assessment of homogeneity.
Similar experiments are carried out for every candidate
reference material of the BCR. Similar findings have been
obtained especially for the two other BCR milk powder samples:
BCR nrs 63 and 150.

Table 3 strictly proves a good homogeneity for the elements
Br, Cu, Na, Fe, Rb and Zn. Experience has shown, however,
that generally when different elements like Zn, Fe, Br and Cu
are homogeneously distributed, the other elements also show a
sufficient homogeneity. The assumption of good homogeneity for
the other elements should be confirmed by the results obtained
in the certification campaign.

## 4. THE CERTIFICATION OF MINOR CONSTITUENTS IN BCR 63 (SKIM MILK POWDER)

Determinations of major nutritive constituents are frequently
carried out in dairy products. As discrepancies between the
results of these measurements can cause problems in trade and
use, the Community Bureau of Reference (BCR) has decided to
certify the contents of $N_{Kjeldahl}$, P, Cl, Na, K, Mg, and Ca in
a skim milk powder.

The range of matrices of dairy products is too large to
produce (stable and homogeneous) reference materials for all
these matrices. Skim milk powder was considered to be a good
starting material, which could serve already to help solving
many problems occurring in dairy quality control.

Before starting with the certification work the homogeneity
and the stability of the sample were investigated. The
homogeneity was tested in a similar way as described in
section 3. The stability under bottling conditions was
studied by analysing the sample over a few years. Table 4
shows that the observed changes from 1978 till 1983 fall
within the analytical errors. This study was undertaken by
one laboratory which principally applied the same method over
these five years.

The certification campaign involved (as pointed out before)
various techniques of fully differing principles and involved
various laboratories to apply these techniques. There was one
exception. The Kjelhahl nitrogen content was determined,
again by various laboratories, using one technique only. In
this case the value depends on an experimental parameter and
can only be measured using a well defined experimental
procedure. The laboratories all follow strictly the
prescribed procedure.

TABLE 4 :   Results over 5 years

| Component | Found in 1978 | Found in 1980 | Obtained as a result of the 1983 study | |
|---|---|---|---|---|
| | | | Value | Standard deviation |
| Mass fractions expressed as : mg g$^{-1}$ | | | | |
| N (Kjeldahl) | 59.36 | 58.86 | 58.66 | 0.52 |
| P | 10.07 | 10.15 | 10.51 | 0.67 |
| Cl | 10.38 | 10.57 | 10.71 | 0.56 |
| Ca | 12.66 | 12.79 | 12.56 | 0.72 |
| K | 18.06 | 17.75 | 17.84 | 1.29 |
| Na | 4.67 | 4.44 | 4.55 | 0.26 |

It would lead too far to present in this context a detailed
description of the way of certification for all the elements.
The element calcium was chosen as a typical example.

Table 5 presents a summary of the methods applied by the
several laboratories.  Widely differing techniques of final
measurement were applied: e.g. flame emission spectrometry
(FES), flame atomic absorption spectrometry (AAS),
chelatometric titration (COMT), ion chromatography (IC),
gravimetry (GRAV), inductively coupled plasma emission
spectrometry (ICP), direct potentiometry with an ion sensitive
electrode in various modes (ISE) and γ-emission spectrometry
(after neutron activation, NAA).

These final measurements were combined with several
pretreatment techniques like: oxygen flask combustion,
dispersion in water, dry ashing and wet digestion with various
reagents.

Figure 2 presents a summary of the findings.  The length of
the bars corresponds with the 95%-confidence interval of
that value obtained in a laboratory.  The abbreviations stand
for the various techniques of final measurement (see above).

For those who have participated often in intercomparisons, the
agreement between the results is striking.

The precision of the various techniques is different.  But in
general, the precision of a certain technique is approximately
the same in all the laboratories applying this technique.
This indicates that the laboratories have applied the
techniques at a good reproducible level of performance.  As
said before, this is a prerequisite for certification.  The
only exception to this statement is one of the direct
potentiometric (ISE) measurements.  The large confidence
interval in one of the ISE-cases is caused by the special way
of application of the technique in which accuracy was

TABLE 5 SUMMARY OF TECHNIQUES AND SAMPLE MASSES AS APPLIED IN THE

DETERMINATIONS OF CALCIUM

| ELE-MENT | SAMPLE MASSES (mg) | SAMPLE PRETREATMENT | FINAL DETERMINATION |
|---|---|---|---|
| Ca | 1000 | Diss. in water (40-50 °C); filtration; add. of NaOH and EDTA | FES |
| | 100 | Pressurised digestion with $HNO_3/HClO_4$ (6 h at 160 °C); add. of Cs (I) | Flame AAS |
| | 1000 | Diss. in water (40-50 °C); filtration; add. of NaOH and EDTA | FES |
| | 1000 | Diss. in water (40-50 °C); filtration | Chelatometric titration |
| | 50 - 100 | Dry ashing at 450 °C; diss. in $H_2O$ | IC |
| | 500 | Dry ashing at 450 °C; diss. in $H_2O$; add. of oxalate | Gravimetry |
| | 1000 | Dry ashing at 450 °C followed by wet ashing; addition of oxalate | Gravimetry |
| | 1500 | Wet digestion with $HNO_3$ | ICP |
| | 400 - 600 | Dry or wet ashing | Flame AAS |
| | 400 - 600 | Dry ashing till 520-530 °C; neutralised soln. | ISE using Gran's plot |
| | | Dry ashing till 520-530 °C; neutralised soln. | ISE |
| | 500 | Diss. is 500 ml water; add. of La (III) 6. | Flame AAS |
| | 500 | Neutron irradiation during 15 sec.; decay time 300 sec. | $\gamma$-counting of $^{49}Ca$ |
| | 500 | Neutron irradiation during 2h in rotating rig; decay time 8 days | $\gamma$-counting of $^{47}Sc$ |
| | 50 | Oxygen flask combustion; abs in $H_2O/H_2O_2$ | ISE |
| | 500 | Neutron irradiation 15 min | $\gamma$-counting of $^{49}Ca$ |
| | 100 | Wet ashing | Flame AAS |
| | 15 | Oxygen flask combustion; pH 13.4 | Chelatometric titration |
| | 1000 | Diss. in water (40-50 °C); filtration | Chelatometric titration |
| | 200 | Digestion with $HNO_3/HClO_4$ | ICP |
| | 200 | Dry ashing at 550 °C; diss. in dil. HCl; add. of La (III) | Flame AAS |
| | 500 | Neutron irradiation 5 min | $\gamma$-counting of $^{49}Ca$ |
| | 1000 - 1500 | Dry ashing; neutron irradiation 5 min; radiochemical separation of $^{49}Ca$ as Ca-oxalate | $\gamma$-counting of $^{49}Ca$ |
| | 700 | Neutron irradiation 7h; decay time 1 week | $\gamma$-counting of $^{47}Sc$ |

RM 63 CALCIUM mg.g⁻¹

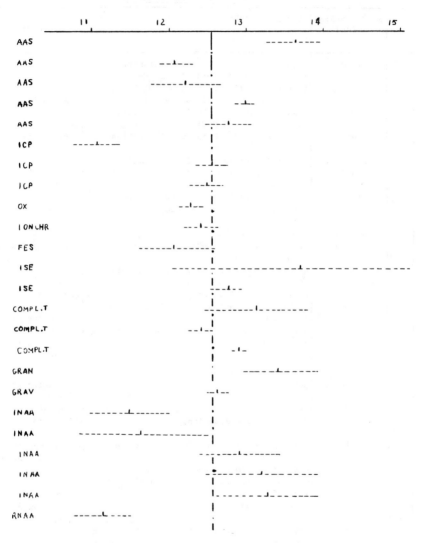

Figure 2

considered to be more important than precision. A similar
explanation can be given for the differences in length of the
confidence interval between one of the chelatometric titration
results and the others.

The agreement between the various results is sufficient. No
bias can be detected between the results of one technique and
the other technique.

A technical discussion between all the participants has shown
that none of the techniques, in the way they were applied,
could be rejected because of a doubt, possible bias etc.

This statement in turn means that statistics can be applied
using the mean values of a set of results given by one
particular laboratory. The statistical evaluation of the data
involved minimally: a detection of outlying variances
(Cochran), of outlying mean values of sets (Dixon) and a test
for the normality of the population (Kolmogorov). The
homogeneity of the variances was verified with the Bartlett
test.

The results which are acceptable after a careful technical and
statistical examination, now can underly the certification.
The certified values are the arithmetic means of the means of
the various sets which have passed the evaluation. Together
with the means the 95%-confidence intervals are calculated.

This was done similarly for the components to be certified,
with the exception of the Kjeldahl-nitrogen content. This
value is defined experimentally. Therefore, the certifying
laboratories had to follow closely the well defined procedure.
The participants were actively participating in the IDF-
working group on the Kjeldahl nitrogen determination and so
they were well experienced in the analytical procedure to be
followed and in the particular type of matrix.

Figure 3 presents a graphic summary of the findings. First of
all, the difference in width of the confidence interval is
striking. Standard deviations varied in the range of 0.05 to
0.39 mg/g. Further on the range of results varies from 57.4
mg/g to 58.8 mg/g. It is clear from figure 3 that although
one can prescribe a method there will remain differences
between laboratories causing variations in standard deviation.

The descrepancies between laboratories for Kjeldahl nitrogen
are much larger than the within laboratory standard deviation.
In these conditions, BCR would normally not certify a
reference material. This spread could not be reduced even
with the laboratories specializing in this field. It was
decided to consider the mean value as a certified one because
the spread of results is nevertheless small and the
uncertainty resulting from the values is sufficient to allow
the material's use as a certified reference material.

Daily practice in the dairy institutes urgently needs a
material whith known Kjeldahl nitrogen content to control the
quality of these determinations. The opinion of the consulted
institutes was that the value already obtained by BCR could
be of good help for analytical quality control.

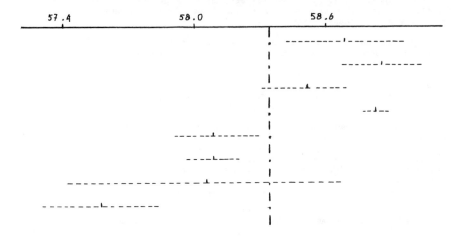

Figure 3

So, finally the decision was made to certify the Kjeldahl-
nitrogen content.

Table 6 finally presents the contents certified in BCR nr 63.
The material is available at the BCR by April 1984.  It will
be delivered together with a certificate and a report which
describes the various aspects of the certification work and
which contains all the individual results.


5.  WORK TO CERTIFY SOME TRACE ELEMENTS IN MILK POWDERS

Trace element analysis at below $\mu g\ g^{-1}$ levels even in
relatively simple matrices poses a considerable problem.  Many
intercomparisons have shown that analytical agreement is
difficult to achieve.  Therefore, e.g. within the IDF, working
groups have worked for many years in order to obtain generally
accepted standardised methods.  The heavy metals working group
has published methods for copper and iron.  For other elements
of concern like mercury, lead and cadmium no standardised IDF
method exists yet.  The very low contents in normal milk make
the determination of these elements difficult; losses occur
easily and contamination is difficult to avoid.  So, a
reference material to improve the situation was highly desired
in 1981.  Such a reference material could in addition be used
to control the already existing methods for copper and iron.

TABLE 6 MAJOR COMPONENTS IN A SKIM MILK POWDER

| Component | Mass fraction (based on dry mass) | | Number of accepted sets of results p |
|---|---|---|---|
| | Certified value([1]) expressed as mg g$^{-1}$ | 95 % confidence interval([2]) expressed as mg g$^{-1}$ | |
| Ca | 12.6 | ± 0.3 | 24 |
| Cl | 10.7 | ± 0.3 | 20 |
| K | 17.8 | ± 0.7 | 14 |
| Mg | 1.12 | ± 0.03 | 13 |
| N (tot.) | 58.8 | ± 0.9 | 9 |
| N (Kjeld.) | 58.5 | ± 0.4 | 6 |
| Na | 4.57 | ± 0.16 | 11 |
| P | 10.4 | ± 0.3 | 12 |

([1]) This value is the arithmetic mean of p values, each value being the mean of a set of 3 to 7 results as provided by the laboratories participating in the certification, expressed on the basis of dried material.

([2]) The 95 % confidence interval is a measure of the uncertainty and is applicable when the reference material is used for calibration purposes. When the reference material is used to assess the performance of a method, the user should refer to the recommendations laid down in the last chapter (instructions for use) of the certification report.

A low level material is not always useful from an economical point of view. Often an r.m. with values in the order of magnitude of the maximum allowable concentration is of more importance. So, in 1981 it was decided to work on the natural milk powder BCR nr 63 and to create two candidate reference materials both spiked with heavy metals. The contents of these two materials (BCR nrs 150 and 151) are resp. below and above the allowable concentrations.

After their preparation their homogeneity was checked (see section 3).

Although the certification work is not yet terminated, some interesting preliminary results can already be described.

A first example deals with the cadmium content of BCR 63.

Figure 4 presents a survey of the findings. The content being in the order of magnitude of 1-10 ng g$^{-1}$ only sensitive techniques could be applied by well experienced laboratories under clean bench or clean room conditions. Techniques of (final) determination were: electrothermal atomic absorption spectrometry (ETAAS), differential pulse anodic stripping voltammetry (DPASV) and isotope dilution mass spectrometry (IDMS). The sample pretreatment consisted mostly of closed vessel or bomb digestion with oxidising acids.

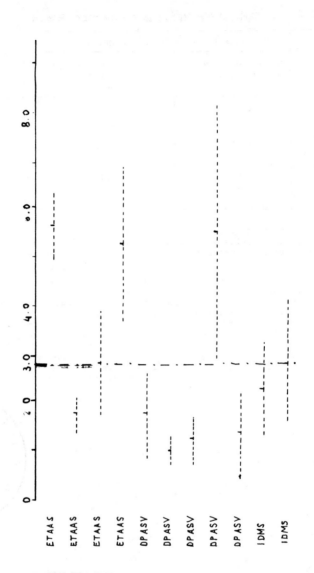

MILK POWDER 63 Cd ng.g⁻¹

Figure 4

Obviously, many laboratories worked at their lower limits.
Therefore, it can not be expected that the same methods led to
a same degree of uncertainty. The various results point at a
value of about 3 ng g$^{-1}$.

The general agreement between the laboratories for a cadmium
determination at such a low level is a good achievement of
this exercise.

A less difficult element and already present at a somewhat
higher level in BCR 63 (about 100 ng g$^{-1}$) is lead. The
results of the determinations of lead in BCR 63 are presented
in figure 5. Spectrometry (SPEC) is applied besides the
already previously mentioned techniques like: ETAAS, DPASV and
IDMS. The agreement between the various methods is excellent
and (with the exception of one outlying variance), the
variances of the various results do not differ too much. For
the determination of copper in BCR nr 63 (fig. 6) more
techniques could be applied. Besides, ETAAS, DPASV, SPEC and
IDMS, inductively coupled plasma (ICP) and other plasma
emission techniques (PES), instrumental neutron activation
analysis (INAA) and flame atomic absorption spectrometry (AAS)
were applied with sufficient agreement. No methodological
bias could be detected between the various methods or between
one method and the whole population of the other methods.

Figure 7 presents the results obained in the determination of
copper in the spiked sample BCR nr 151. Besides the methods
used for the determination of copper in BCR nr 63,
potentiometric stripping analysis (POTST) and neutron
activation analysis with radiochemical separation (RNAA) have
been applied. Again, the agreement is good.

Figure 8 finally gives the example of the determination of
iron in spiked sample BCR nr 10. Three variances (AAS, RNAA
and INAA) are clearly outlying as indicated in the figure.
The difference in standard deviation of the sets of results
obtained by the same method of final determination is rather
larger. It is hardly possible to contribute this spread to
the difference in sample pretreatment technique. A bias
between results of the various techniques could not be
detected. Obviously, some laboratories have to improve their
performance.

These examples generally show a good agreement between the
results of laboratories. Determining concentrations in the
order of ng/g up to 100 ng/g is not an easy task. The good
agreement between the results shows that it is possible to
perform trace analysis at this level.

The agreement obtained here is much better than can be
expected in a normal intercomparison. Clearly, a careful
selection of the laboratories invited to participate in a
certification campaign and the special care these laboratories
have given has been necessary to obtain this result.

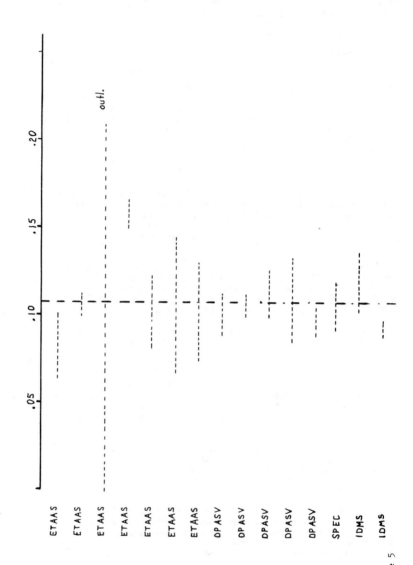

Figure 5

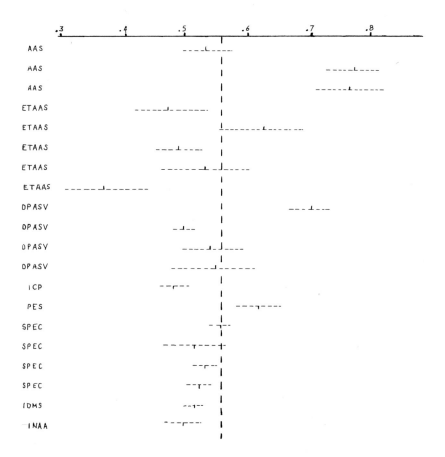

Figure 6

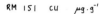

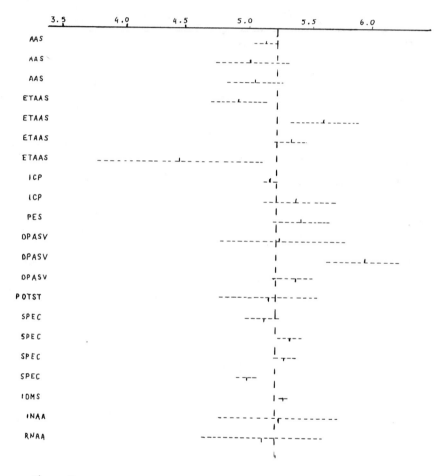

Figure 7

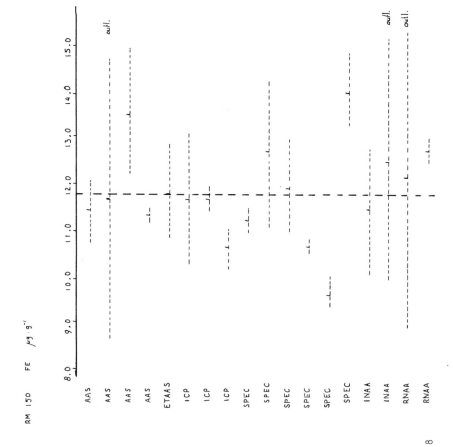

Figure 8

For a decision about certification quite a few considerations
have to be taken into consideration:

i)    do the results of various principally differing methods
      agree sufficiently,

ii)   can it be expected that all participants have been
      working according to the best state of the art,

iii)  is there a methodological bias,

iv)   is the figure to be certified accurate and of sufficient
      precision for the expected use in practice.

The examples shown above illustrate several situations which
can be expected in evaluating analytical results.

The work on the trace element contents in the three milk
powder samples will probably be concluded before June 1984.
It can be expected that the materials will become available in
October 1984.

# Results of a FECS/WPFC Cooperative Study on the Determination of Aflatoxin $M_1$ at Low Levels in Dried Milk

By R. Battaglia[1]*, H. P. van Egmond[2], and P. L. Schuller[2]
[1]CANTONAL LABORATORY, P.O. BOX, CH-8030 ZÜRICH, SWITZERLAND
[2]INSTITUTE OF PUBLIC HEALTH AND ENVIRONMENTAL HYGIENE, BILTHOVEN, THE NETHERLANDS

Introduction

Due to the highly suspected carcinogenic potential of aflatoxin $M_1$ and the high consumption of dairy products, the statutory limits of this food-contaminant are generally very low in the mg/kg or even ng/kg-range. The lowest limit so far has been set by Switzerland, where whole milk must not contain more than 50 ng/kg; if it is intended for baby-food it must not contain more than 10 ng $AFM_1$/kg.

There is, at the moment, no generally recognised analytical method available, which allows to determine aflatoxin $M_1$ at these low levels with an acceptable precision ("generally recognised" would also mean: collaboratively tested!). The consensus reached by the Mycotoxin-Group of the IUPAC-Food Commission in 1982 was, that the chances of having such a method at hand in the near future are extremely slim. However, low statutory limits have been issued and other European countries consider lowering their limits to the ng/kg-region as well. The comparability of analytical results in this field is therefore for obvious reasons of prime importance. It was for this reason mainly, that the Working-Party on Food Chemistry of the FECS decided to conduct a cooperative study within Europe on the determination of aflatoxin $M_1$ at low levels in milk-powder (COLAB I). Since a comparative study in the Netherlands was planned, coordinated by Dr. P.L. Schuller and Ir. H.P. van Egmond of the Laboratory for Chemical Analysis of Foodstuffs of the National Institute of Public Health and Environmental Hygiene, it was agreed that the milk-powder prepared for that study could be used for COLAB I as well.

## Set-up of the test

In a first letter, the national representatives of the FECS-
WPFC (16 countries at that time) were asked to distribute an
invitation to the food-analysts of their countries. A total of
90 laboratories responded and wanted to take part in the study.

It was then made clear to all prospective participants that they would
obtain three samples for analysis, that the target levels were
in the ng/kg-range and that they would be supplied with a practice-sample containing between 200 and 400 ng aflatoxin $M_1$ per kg.
The choice of the method was free. In a brief questionnaire, the
participants were asked to give information on their degree of
experience in this field (low, medium, high), on the methods
used, the recoveries etc.

## Preparation of samples of milk-powder and quality assurance

Seventeen cows of the Research Institute for Animal Feeding and
Nutrition (Lelystad, the Netherlands) were fed grass only during
a seven-day period, as to obtain milk free from aflatoxin $M_1$.
Simultaneously 5 cows of the same type daily received two supple-
mental nutrient briquettes, each artificially contaminated with
an amount of 0.2 mg of aflatoxin $B_1$ as to obtain milk naturally
contaminated with aflatoxin $M_1$. (Analysis of the briquettes (ca.
85 g each) before addition of aflatoxin $B_1$ had shown these were
naturally contaminated  with aflatoxin $B_1$ at a level of ca. 5
$\mu$g/kg). From a carry-over study conducted at the National Institute of Public Health (1), it was known that a daily intake of
0.4 mg of aflatoxin per cow would lead to a concentration of
aflatoxin $M_1$ in the milk ranging from 0.24 - 0.36 $\mu$g/l. The contaminated and uncontaminated milks were collected for three days,
starting at day five, and aflatoxin $M_1$ concentrations were determined in quadruplicate according to the procedure of Stubblefield
(2) with a few modifications (3).

The aflatoxin $M_1$ concentrations in the contaminated and uncontaminated milks were found to be 0.264 $\mu$g/l and non-detectable
(<0.010 $\mu$g/l) respectively. Immediately after collection and
analyses the batches of milk were transported to the Netherlands

Institute for Dairy Research, Ede, the Netherlands. A part of the uncontaminated milk was kept aside, three other parts of the uncontaminated milk were mixed with contaminated milk in ratio's of 96:4, 88:12
and 97:23 respectively, as to obtain batches of milk, contaminated with
aflatoxin $M_1$ at four different levels: 0, ca. 0,01, ca. 0,03 and
ca. 0.05 µg/l. Then these batches of milk were spray-dried according to usual procedures of the Dutch dairy industry. The
milk-powders with estimated levels of aflatoxin $M_1$ of 0, ca. 0.1, ca. 0.3
ca. 0.5 µg/kg respectively were packed in plastic bags and stored
at $-10^\circ$C immediately after production, until sub-packing. Shortly
after production, 40 gram amounts of the milk-powders with estimated $M_1$ concentrations of 0,0.1 and 0.5 µg/kg ("unknown" samples)
and 80 gram amounts of milk-powder with an estimated concentration
of 0.3 µg/kg ("practice" sample) were weighed into laminate pouches
for food use, that were sealed under $N_2$ at Gist Brocades b.v. Delft,
the Netherlands. For Dutch collaborators smaller amounts of milk-
powder (5 or 10 gram) were packed. The "unknown" samples were la-
belled with randomly drawn numbers and together with the practice
samples packed in small carton boxes which were sent to collaborators.

Additionally several experiments were carried out to assure the
quality of the samples, relative to the concentration of aflatoxin
$M_1$. From the batch with an estimated concentration of aflatoxin
$M_1$ of 0 µg/kg three randomly taken pouches were analysed with the
modified procedure of Stubblefield (3), to ensure aflatoxin $M_1$ was
not detectable. The batch with an estimated concentration of aflatoxin $M_1$ of 0.3 µg/kg was taken to check the homogeneous distribution of aflatoxin $M_1$ in milk-powders obtained in the way described. For this purpose 24 samples taken out of ca. 300 were analysed with the modified method of Stubblefield (3). In addition
samples of this batch were sent to three other Dutch Institutions
experienced in the field of aflatoxin $M_1$ analysis (Food Inspection
Service, Rotterdam; State Institute for Quality Control of Agricultural Products, Wageningen; the Netherlands Controlling Authority
for Milk and Milk Products, Leusden), as to calculate a "best
value" for the actual aflatoxin $M_1$ concentration in the practice
sample.

The analyses revealed that aflatoxin $M_1$ was not detectable
($<0.1 \mu g/kg$) in the batch with the assumed aflatoxin $M_1$ concentration of $0 \mu g/kg$ and that aflatoxin $M_1$ was homogeneously distributed ($\bar{x}$ = $0.29 \mu g/kg$, (not corrected for recovery) c.v. = 20 %)
in the batch with an estimated aflatoxin $M_1$ concentration of 0.3
$\mu g/kg$. The "best value" ranged from $0.2 - 0.4 \mu g$ $M_1/kg$.

Based upon these findings it was concluded that the materials
were suitable for the collaborative study.

## Results and Discussion

A total of 58 laboratories sent in results. The individual analytical data and some of the information given in the questionnaire are tabulated in Table 1. Further information on the various practices of handling and assaying of standards, the use of
TLC and HPLC, the absolute detection-limits etc. can be found in
Table 2. Since a wide variety of methods has been employed,
no attempt to compare methods by means of statistical analysis
was envisaged. However, attention was given to a differentiated
presentation of the results. For all presented frequency-distri-
butions, all $\bar{x}$-values from Table 1 except the "not detectables"
($\bar{x}$ =0) have been considered. As explained in the footnotes to
Table 1 , these final results ($\bar{x}$) were obtained by averaging the
individual results (except those marked as "outliers" by the participants, *i.e.* values in brackets) and correcting them for
the cited recoveries. Based on the answers to the question "how
many samples did you analyse in 1982? (1(0) - 10, 11 - 100,  100)",
the data  were grouped in classes of "low", "medium" and "high"
experience.

## Sample "0"

As explained earlier, the probability that this sample contained
any aflatoxin at all is extremely low, since it was prepared from
milk of cows feeding on fresh grass only. Therefore, the positive
results must probably ·be termed as false positives. Figure 1 shows
the frequency-distribution of these results. 18 participants detected aflatoxin $M_1$ in this sample. These positive findings are
compared with the degree of experience of the participants in Table 3.

Table 1

Explanation

Nr     = number of participant

Ex     = experience-class (1 for "low", 2 for "medium", 3 for "high")

Method = cited as mentioned by the participant. *:L.G.M.Th.Tuinstra,
         W.Haasnoot, Z.Anal. Chemie 312, 622 (1982), adapted for TLC

Rec    = recovery,in %,where determined or cited.

x̄      = mean of individual results except figures in brackets (marked
         as outliers by the participant), corrected with recovery where
         possible.

Table 1

| Nr | Ex | Method | Rec | Sample"0" individ.results | x̄ | Sample"0.1" individ.results | x̄ | Sample"0.5" individ.results | x̄ |
|----|----|--------|-----|---------------------------|-----|------------------------------|-----|-----------------------------|-----|
| 1 | 2 | Schuller(TLC), modified | – | (20),29,44,12 | 28 | (69),135,103 | 119 | (179),234,278, 253 | 255 |
| 2 | 1 | * | 85 | 0 | 0 | 10 | 12 | 300,310 | 359 |
| 3 | 1 | * | 85 | 0 | 0 | 100 | 118 | 400,290 | 406 |
| 4 | 3 | * | 85 | <100 | 0 | 100 | 118 | 150, 100 | 176 |
| 5 | 1 | * | 85 | 200 | 235 | 100 | 118 | 350,300 | 382 |
| 6 | 1 | * | 85 | 0 | 0 | 25 | 29 | 200,200 | 235 |
| 7 | 1 | * | 85 | 300 | 353 | 100 | 118 | 150,100 | 147 |
| 8 | 1 | Stubblefield, modified | 99 | 0 | 0 | 0 | 0 | 120 | 120 |
| 9 | 1 | * | 85 | n.d. | 0 | n.d. | 0 | 50,100 | 88 |
| 10 | 1 | * | 85 | <250 | 0 | <250 | 0 | 400,500 | 529 |
| 11 | 1 | * | 85 | <100 | 0 | <100 | 0 | 400,150 | 324 |
| 12 | 2 | * | 85 | n.d. | 0 | 162 | 191 | 503,401 | 532 |
| 13 | 2 | * | 85 | n.d. | 0 | n.d. | 0 | 750,750 | 882 |
| 14 | 1 | * | 85 | 100 | 118 | <100 | 0 | 300,250 | 324 |
| 15 | 2 | * | 85 | <100 | 0 | <100 | 0 | 100,200 | 176 |
| 16 | 1 | * | 85 | n.d. | 0 | n.d. | 0 | 375,250 | 368 |
| 17 | 1 | * | 85 | 500 | 588 | <100 | 0 | 250,250 | 294 |
| 18 | 1 | * | 85 | 120 | 141 | n.d. | 0 | 500,400 | 529 |
| 19 | 1 | * | 85 | n.d. | 0 | n.d. | 0 | 500,500 | 588 |

Table 1 cont.

| Nr | Ex | Method | Rec | Sample"0" | | Sample"0.1" | | Sample"0.5" | |
|----|----|--------|-----|-----------|-----|-------------|-----|-------------|-----|
| | | | | individ.results | x̄ | individ.results | x̄ | individ.results | x̄ |
| 20 | 2 | * | 85 | n.d. | 0 | 120 | 141 | 280,430 | 424 |
| 21 | 2 | * | 85 | <50 | 0 | 80 | 94 | 250,150 | 235 |
| 22 | 2 | * | 85 | n.d. | 0 | 90 | 106 | 370,410 | 459 |
| 23 | 3 | Leuenberger et al. J.Chromatog.178,543 (1979) | 70 | 22,26 | 34 | 171,175 | 247 | 330,302 | 451 |
| 24 | 1 | Schuller, as published | 90 | 0,0,0,0 | 0 | 0,0,0 | 0 | 250,300 | 306 |
| 25 | 1 | Guggisberg,modified | – | 65,43 | 54 | 163,173 | 168 | 477,412 | 445 |
| 26 | 1 | (?) no method cited | – | 0,0,0 | 0 | 133,50 | 92 | 210,250,274 | 245 |
| 27 | 1 | own method | 85 | <50,<50,<50 | 0 | 66,89 | 92 | 357,341,390 | 440 |
| 28 | 3 | AOAC | 67 | <30,<30,<30 | 0 | 77,51,45 | 87 | 208,192,291 | 358 |
| 29 | 2 | stubblefield,modified | – | 150,120 | 135 | 90,110 | 100 | 250,280 | 265 |
| 30 | 1 | L.Tuinstra et al., J.Chromatog. 111,448 (1979) | – | 186,194,185 | 188 | 282,297,310 | 296 | 487,579,525 | 530 |
| 31 | 2 | Leuenberger,modified | 72 | 27,58,67 | 50 | 56,96,50 | 67 | 270,330,290 | 296 |
| 32 | 1 | Guggisberg,as supplied | 70 | <100,<100,<100 | 0 | <100,70 | 0 | 230,170,170 | 271 |
| 33 | 3 | "Amtliche Methoden-Sammlung" BRD June 1982 | 80 | 368,276,338 | 327 | 324,396,480 | 400 | 660,612,740 | 671 |
| 34 | 1 | Leuenberger,modified | 80 | n.d.,n.d.,n.d. | 0 | 75,65,80 | 91 | 485,465,460 | 588 |
| 35 | 2 | J.AOAC 63, 394 (1980), modified | 80 | 32,24,25 | 33 | 91,71 | 101 | 402,375,316 279,368 | 440 |

Table 1 cont.

| Nr | Ex | Method | Rec | Sample"0" individ.results | x̄ | Sample"0.1" individ.results | x̄ | Sample"0.5" individ.results | x̄ |
|---|---|---|---|---|---|---|---|---|---|
| 36 | 1 | Guggisberg,modified | 62 | <20, <20, <20 | 0 | 90,97,63 | 83 | 379,399,455 | 411 |
| 37 | 1 | "Amtliche Methoden Sammlung"BRD | | <200,<200,<200 / <200,<200, 269 | 0 | <200,<200,<200 / <200,<200, 331 | 0 | 438,345,410 / 407,372 317 | 381 |
| 38 | 2 | own method | 82 | < 50, <50,< 50 / <50, <50,< 50 | 0 | 100,66,74 / 84,108,98 | 107 | 339,299,298 / 303,342,(172) | 384 |
| 39 | 3 | Bundesgesundheits-blatt 23 (1980), modified | 90 | (46), 0 | 0 | 104,116,(54) | 110 | 441,550,500 | 497 |
| 40 | 2 | Guggisberg, modified | – | 9,9,9,(n.d.) | 9 | 76,74,79,72 | 75 | 356,364,336,407 | 352 |
| 41 | 2 | Guggisberg,as supplied | 105 | n.d.,n.d. | 0 | n.d.,n.d. | 0 | 30,50 | 40 |
| 42 | 1 | own method | 95 | n.d.,n.d. | 0 | 77,78 | 77 | 247,248 | 247 |
| 43 | 2 | Guggisberg,modified | 85 | <40,<40,<40 / <40 | 0 | 82,64,53 / 83,67 | 82 | 383,400,415 / 360,339 | 446 |
| 44 | 2 | Leuenberger,modified | – | 80,80,80 | 80 | 120,117,120 | 119 | 300,300,280 | 293 |
| 45 | 2 | Kiermeier et al.,ZLUF 160,337 (1976),mod. | 95 | (30),(48);comment:"free of $AFM_1$" | 0 | 168,147 | 166 | 375,387 | 401 |
| 46 | 1 | Guggisberg,modified | 64 | 36,38 | 37 | 31,41,41 | 38 | 80,83,73 | 79 |
| 47 | 1 | Steiner et al., as supplied | – | 0,0,0 | 0 | 50,34,38 | 41 | 150,97,200 | 149 |
| 48 | 2 | Stubblefield, modified | 90 | 0,0,0 | 0 | 81,75,48 | 76 | 280,276,193 | 278 |
| 49 | 3 | own method | 70 | <170,<110,<150 | 0 | <220,<70,<80 | 0 | 320,400,180 | 300 |

Table 1 cont.

| Nr | Ex | Method | Rec | Sample"0" | | Sample"0.1" | | Sample"0.5" | |
|----|----|--------|-----|-----------|---|-------------|---|-------------|---|
| | | | | individ.results | x̄ | individ.results | x̄ | individ.results | x̄ |
| 50 | 3 | Guggisberg,as sup-plied | 85 | 14,23,11,12 | 18 | 87,96,75,90 | 102 | 397,457,446,393,402 | 493 |
| 51 | 3 | Guggisberg,as sup-plied | 90 | n.d.,n.d.,n.d., | 0 | 86,93,59 | 82 | 357,355,352 | 394 |
| 52 | 3 | Leuenberger et al., J. Chromatog.178, 543 (1979) | 78 | <50,<50 | 0 | 123,136 | 130 | 548,505 | 526 |
| 53 | 2 | own method | 95 | <50,<50,<50 | 0 | 113,113,116 (98) | 114 | 445,406,368, (330) | 406 |
| 54 | 3 | own method | 95 | <50,<50,<50 <50 | 0 | **133[99],58[74] 106[150],109 [130] | 113 | **473[504],328 [488], 325[470],452 [490], 490[460] | 471 |
| 55 | 2 | Leuenberger, mod. | 75 | 55,(73),44,51 | 67 | 78,(117),78,84 | 107 | 312,(351),318, 270 | 400 |
| 56 | 3 | own method | 77 | n.d., n.d. | 0 | 113,109,110, 110 | 110 | 331,298,306 339 | 319 |
| 57 | 2 | AOAC, modified | – | 0,0,0 | 0 | 0,0,0 | 0 | 0,0,0 | 0 |
| 58 | 2 | AOAC | – | 0,0,0 | 0 | 0,0,0 | 0 | 0,0,0 | 0 |

**values in []: derivatised samples

Table 2

Compilation of answers in the questionnaire

| Query | Answer |
|-------|--------|
| Calibration of Standards<br>    AOAC-procedure<br>    supplier<br>    check-samples (e.g.EEC) | 25 participants<br>24      "<br>3       " |
| Quantitation<br>    TLC/densitometry<br>    TLC/visual estimation<br>    HPLC/fluorescence | <br>36 participants<br>19      "<br>3       " |
| Detection-limits (S/N=3)<br>    0.01 - 0.09 ng<br>    0.09 - 0.17 "<br>    0.17 - 0.25 "<br>    0.25 - 0.33 "<br>    0.49 - 0.57 "<br>        1.0     " | <br>19 participants<br>5       "<br>5       "<br>2       "<br>3       "<br>1       " |
| Methods used<br>    "exactly as published"<br>    own methods,unpublished<br>    modified methods<br>    no reply | <br>15 participants, 7 methods<br>6       "        6      "<br>36      "        17     "<br>1 participant  ———<br>                30 diff.methods! |

Table 3

Results for sample "0"

| Group | no AFM$_1$ detected | positive results | range, ng/kg |
|-------|---------------------|------------------|--------------|
| low experience | 18 | 8 | 37 - 588 |
| medium experience | 14 | 7 | 9 - 135 |
| high experience | 8 | 3 | 18,34,327 |

Sample "0.1"

The results are represented in figures 2 to 5. Since it was decided from the outset of the study, that apparent and obvious
"outliers" or "extreme values" shall not be eliminated, the coefficients of variation alone are of limited meaning; the graphical
representations given in the figures show in an obvious way, that
the scatter of results becomes smaller with increasing experience.
The influence of the experience is quite drastic in the "not detectables"; in the "high experience" class 9 % reported"not detectable",in the "medium experience" class there are 24 % of "not
detectable" results and in the "low experience" class there are
46 %. The conclusion to be drawn therefore seems clear. The  limit
of detection does not so much depend on the choice of the method
but on the experience of the chemist carrying out the analyses. It
has to be remembered here, that each participant had the opportunity to evaluate and choose a method with respect to the required
sensitivity, since the content of the practice-sample (200 - 400
ng/kg) was known.

Sample "0.5"

The results are represented in figures 6 to 9. Here again, a dependence on the degree of experience may be seen. The coefficient
of variation for the results of all laboratories (42.4 %) is in
comparison to other collaborative (4) and cooperative (5) studies
at much higher levels quite low: the corresponding coefficients
of variation were - after elimination of outliers! - 36.2 % and
47.3 % (levels 3000 and 1120 ng/kg, Table 11 in Lit.4), and 44 %
(level 8000 ng/kg, Table 2 in Lit. 5). The coefficient of variation
in the high experience-group  (11 laboratories, 10 different methods) is 31.5 % only. Table 4 gives a further survey of the effect
of experience.

### Table 4
Comparison of "Experience-classes"

|  | "High + Medium exp." | | "Low exp." | |
|---|---|---|---|---|
|  | CV | "n.d." | CV | "n.d." |
| Sample  0.1 | 54 % | 18.8% | 72.2% | 46.2% |
| Sample  0.5 | 40.7% | 6.3% | 44.2% | - |

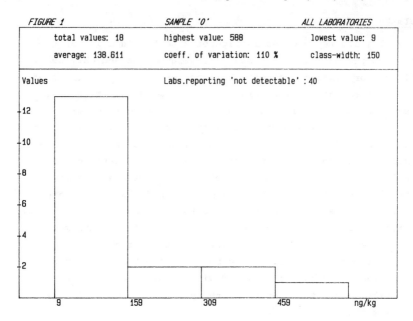

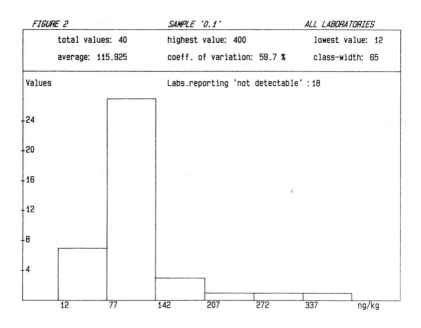

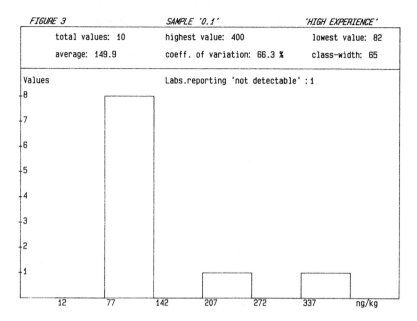

FIGURE 3    SAMPLE '0.1'    'HIGH EXPERIENCE'

total values: 10    highest value: 400    lowest value: 82

average: 149.9    coeff. of variation: 66.3 %    class-width: 65

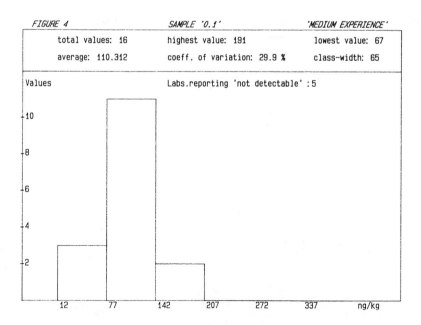

FIGURE 4    SAMPLE '0.1'    'MEDIUM EXPERIENCE'

total values: 16    highest value: 191    lowest value: 67

average: 110.312    coeff. of variation: 29.9 %    class-width: 65

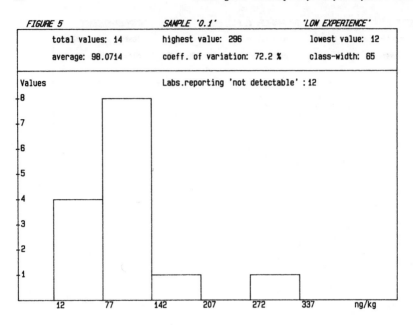

FIGURE 5                    SAMPLE '0.1'                    'LOW EXPERIENCE'

total values: 14          highest value: 296          lowest value: 12

average: 98.0714          coeff. of variation: 72.2 %          class-width: 65

Values                    Labs.reporting 'not detectable' : 12

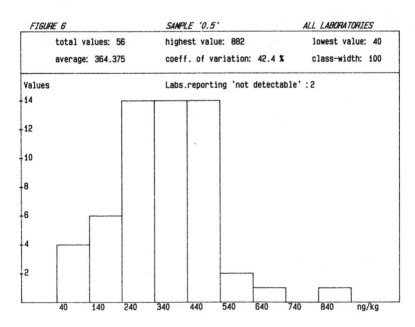

FIGURE 6                    SAMPLE '0.5'                    ALL LABORATORIES

total values: 56          highest value: 882          lowest value: 40

average: 364.375          coeff. of variation: 42.4 %          class-width: 100

Values                    Labs.reporting 'not detectable' : 2

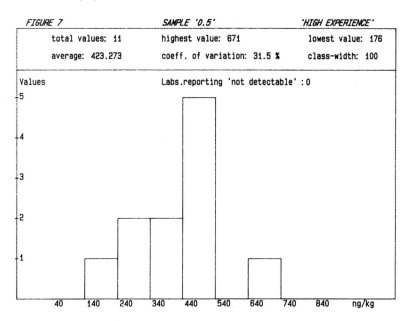

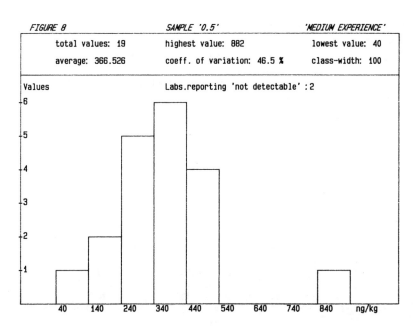

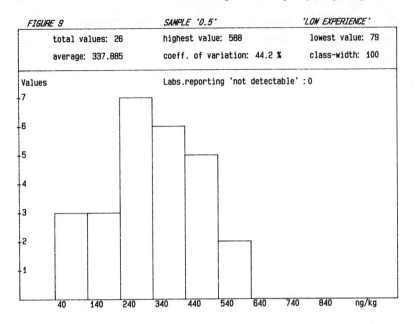

FIGURE 9               SAMPLE '0.5'                    'LOW EXPERIENCE'

total values: 26        highest value: 588        lowest value: 79

average: 337.885        coeff. of variation: 44.2 %        class-width: 100

Values                  Labs.reporting 'not detectable' : 0

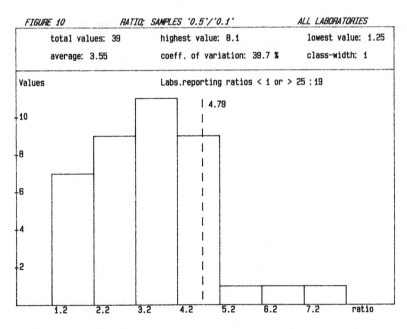

FIGURE 10           RATIO: SAMPLES '0.5'/'0.1'           ALL LABORATORIES

total values: 39        highest value: 8.1        lowest value: 1.25

average: 3.55           coeff. of variation: 39.7 %        class-width: 1

Values                  Labs.reporting ratios < 1 or > 25 : 19

Ratio of results: Sample "0.5" to sample "0.1"

It can be calculated, that the two samples showed an aflatoxin $M_1$ content in the ratio of exactly 4.79. Eliminating the value from participant No. 2 (ratio 29.9) and of course all the"n.d."s an average ratio of 3.55 (median-value 3.56) with a coefficient of variation of 39.7 % results (figure 10). Although a slight tendency towards the true value for the high-experience group (median 4.1) can be calculated, there seems to be a real problem as to the recoveries achieved at different levels. No doubt this will be a field for future work.

Summary

Three samples of milk-powder, one blank, two naturally contamina-ted with aflatoxin $M_1$ in the ratio of 4.79 : 1 at ng/kg levels as well as a practice-sample containing ca. 300 ng aflatoxin $M_1$ per kg have been analysed in 58 laboratories from 13 European countries. The choice of the method was free. The results com-pare very favourably with similar studies at higher aflatoxin $M_1$ levels.

The occurrence of positive results for the blank sample (31 %), false negative results for the contaminated samples (31 %, sample 0.1; 3.4 %, sample 0.5) and the coefficients of variation were lower for the experienced laboratories (54 %, sample 0.1; 40.7 %, sample 0.5) than for unexperienced ones (72.2 %, sample 0.1; 44.2 %, sample 0.5), seemingly independent of the methods employed.

Literature

1) Verhülsdonk, C.A.H.; Paulsch, W.E.; Wierda, W.and Gansewinkel,B.J.van(1972) RIV internal report 137/72 LCLO (1972)

2) Stubblefield, R.D. J.Am.Oil Chem.Soc. 56, 800-802 (1979)

3) Egmond, H.P.van; Paulsch, W.E.; Sizoo, E.A. and Schuller, P.L.

4) R.D.Stubblefield, H.P. von Egmond, W.E. Paulsch, P.L. Schuller, J. AOAC 63.907 (1980)

5) M.D.Friesen, L. Garren, J. AOAC 65, 864 (1982)

List of participants

Austria

R. Pfleger, Vet.-med.University, Vienna

Belgium

J. Bijl, University of Gent, Gent

Denmark

H. Werner, Danish Govt.Res.Inst.for Dairy Ind., Hillerød

Federal Republic of Germany

B. Bergner, Chem.Untersuchungsamt, Stuttgart
Eyrich, Chem. Landesuntersuchungsanstalt, Karlsruhe
H. Gaumnitz, Chem.Landesuntersuchungsanstalt, Freiburg
Hennig, Chem. Landesuntersuchungsanstalt, Münster
A. Horvath, Chem. Untersuchungsanstalt, Pforzheim
G. Josst, Chem. Untersuchungsanstalt, Düsseldorf
Klein, Chem.Untersuchungsanstalt, Aachen
Rohrdanz, Staatl.Chem. Untersuchungsamt, Lüneburg
Stauff, Chem.Untersuchungsamt, Paderborn
V. Heyer, Chem. Untersuchungsstelle der Bundeswehr, Kiel-Kronshagen
Chemisches Untersuchungsamt, Recklinghausen
El-Dessouki, Chem. Landes-Untersuchungsanstalt, Stuttgart
H. Sommer, Chem.Untersuchungsamt, Bielefeld
R. Woller, Chem.Untersuchungsamt, Trier
R. Weber, Max von Pettenkoferinstitut, Bundesgesundheitsamt,Berlin

Finland

E. Lindfors, National Vet. Inst., Helsinki

France

M. Jemmali, IMSERM, Paris
Mazerand, EMSMIC, Paris

Great Britain

M.J. Shepherd, MAFF-Food Laboratory, Norwich
M.V. Howell, RHM Research Ltd., High Wycombe
G.M. Telling, Unilever Res.Lab., Sharnbrook
P.S. Mann, Food Res.Assoc., Leatherhead

Greece

G. Vasilikotis, Aristotelion University, Thessaloniki

Italy

A. Bottalico, CNR Inst.of Toxins and Mycotoxins, Bari
A. Viscontini, "    "    "    "    "    "        Bari

## Netherlands

R. ten Broeke, Food Insp.Service, Amsterdam
J.A.P. Smit, Food Insp. Service, Assen
C.A. Kan, Inst.for Poultry Research, Beekbergen
H.P.van Egmond,National Inst. of Public Health, Bilthoven
H.H. Simons, Coberco Research Laboratory, Deventer
J.v.d. Linden, Food Insp. Service, Dordrecht
J. de Vries, Food Insp. Service, Enschede
A. de Bruin, Food Insp. Service, Goes
J.W.Th. Coolsma, Food Insp. Service,'s-Gravenhage
L.J. Poortvliet, Butter and Cheese Control Station, Leeuwarden
J.A. Jans, Netherl.Contr.Auth.for Milk & Milk Products, Leusden
G.H. Tjan, Food Inspection Service, Nijmegen
J.R. Besling, Food Insp.Service, Rotterdam
G.J. van den Bosch, Coop. Central Lab., Veghel
R.W. Maeyer, Nestlé NL. Regional Lab., Vlaardingen
L.G.M.Th. Tuinstra, RIKILT, Wageningen
E.J. Mulders, Dept. of Tox. Analysis TNO, Zeist
J. Bonnema, Food Insp.Service, Zutphen
Food Insp. Service, Utrecht
H. v.d. Horst, Food Insp. Service, Haarlem
G. van Gestel, Mars Chocolate Fact., Veghel

## Norway

M. Yndestad, Vet. College of Norway, Oslo

## Switzerland

A. Cominoli, Cantonal Laboratory, Oslo
U.P. Buxtorf, Cantonal Laboratory, Basle
A. Camenzind, Gemeinschaftslaboratorium, Belp
H. Guggisberg, Cantonal Laboratory, Frauenfeld
U. Leuenberger, Cantonal Laboratory, Berne
C. Wyss, Nestlé, La-Tour-de-Peilz
W. Steiner, Cantonal Laboratory, Zurich
J. Mitiska, Cantonal Laboratory, Zurich

## Acknowledgements

The authors express their gratitude to Ir.K. Vreman (Research Institute for Animal Feeding and Nutrition, Lelystad, the Netherlands) for preparing milk, naturally contaminated with aflatoxin M₁, to Drs.H.A. Veringa (Netherlands Institute for Dairy Research, Ede, the Netherlands) for processing of the milk to milkpowder and to Ir.A.Zwijgers (Gist Brocades b.v., Delft, the Netherlands) for supplying laminate pouches and for offering facilities to seal the pouches in an appropriate way.

# Collaborative Study on a Potential EEC Method for the Determination of Milk Fat in Cocoa and Chocolate Products

By R. Wood

MINISTRY OF AGRICULTURE, FISHERIES AND FOOD, 65 ROMNEY STREET, LONDON SWIP 3RD, U.K.

SUMMARY

The results of a collaborative study to evaluate a GLC method of analysis to determine milk fat in chocolate are presented. The method is based on the direct determination of butyric acid in extracted fat. The results of the trial are considered to be satisfactory and the method may, therefore, be proposed for inclusion in legislation.

INTRODUCTION

The European Community (EC) Council of Ministers has adopted a Directive relating to cocoa and chocolate products intended for human consumption[1]. This Directive was required to be translated into the national legislation of each of the member states. In the case of England and Wales this was enacted in "The Cocoa and Chocolate Regulations 1976"[2].

Amongst the criteria which are prescribed in the Council Directive are requirements relating to the concentrations of milk fat in certain products, e.g.

at least 3.5% milk fat in milk chocolate

at least 5.0% milk fat in milk chocolate with high milk content.

at least 3.0% milk fat in milk chocolate vermicelli and flakes

at least 3.5% milk fat in white chocolate

There is also a requirement in the Directive that methods of analysis be developed to enforce the provisions in the Directive.

The Community will therefore, at some stage, adopt a method of analysis to determine the milk fat concentration in the above products.

At the present time the methods of analysis which are available to determine milk fat and which have also been standardised and validated are few, these being mainly the classical distillation and titration procedures.

The inclusion of such procedures  in legislation to enforce the milk fat provisions of the Directive would probably be unacceptable to food analysts, many of whom express preferences for a more modern instrumental method to be prescribed by legislation.

Before any such method would be accepted into legislation it would have to be validated, preferably by collaborative trial. The Ministry is able to arrange suitable trials to enable such validation to be achieved.

MINISTRY COLLABORATIVE TRIAL PROGRAMME

The Ministry has sponsored and organised a collaborative trial programme to test or verify methods of analysis for foodstuffs. Such methods are selected for collaborative testing because they have been suggested for incorporation in legislation – either as a result of European Community or national, i.e. purely UK, consideration.  In addition, methods may be selected because they relate to matters of interest to UK food analysts

but will not necessarily be incorporated directly into
legislation.  The determination of milk fat in chocolate is an
example of a potential legislative method.

POTENTIAL METHODS FOR THE DETERMINATION OF MILK FAT
Various categories of methods are available for the determination of milk fat.  These are:

    a:    methods based on semi-micro indices, such as that
        adopted by the OICC[3],

    b:    methods based on gas liquid chromatography (GLC) of
        a specific constituent of milk fat, eg butyric acid
        or cholesterol,

    c:    methods based on triglyceride composition of milk fat.

One of the most commonly used instrumental procedures in the UK
for the determination of milk fat is that proposed by Phillips
and Sanders[4] and which is based on the direct estimation of
butyric acid, and hence milk fat, in the samples, or the
extracted fat from the samples. The GLC conditions for the
method are outlined in Appendix I.

The UK has proposed that this method be adopted by the Community
for the estimation of milk fat in chocolate products.  Two
collaborative trials have been carried out to support that
proposal.

The trials have been based on
    a.   the determination of milk fat in admixtures with

cocoa butter, and

b.   the determination of milk fat in milk chocolate
     samples.

## Milk Fat in  Admixtures with Cocoa Butters Trial

This trial, which took place in 1978, was arranged to test the
method as described by Phillips and Sanders[4] and using the
GLC conditions outlined in Appendix I, and the method as
slightly modified by the Laboratory of the Government Chemist,
on a range of pure fats.

The composition of the 12 samples used in this trial is given
in Table 1.  Samples were sub-sampled to produce both known and
blind duplicates, samples of each code letter being analysed
once only.

Forty analysts took part in the trial (25 UK public analysts,
one Republic of Ireland public analyst, 8 analysts from
industrial or research organisations, 4 from Government
establishments and 2 from Government establishments in the
Netherlands) but of these only 10 used the Phillips and Sanders
method without modification.

Analysts were asked to report as a single result the percentage
by weight of milk fat in the sample as received using a factor
of 3.6% for the conversion of butyric acid to milk fat.

The results of the modified Phillips and Sanders method were
disappointing and that method cannot be recommended.

## Table 1

## Composition of samples used in milk fat in admixtures with cocoa butters trial

| Sample Number | Sample Letter Code | Composition (g/100g) | | | |
|---|---|---|---|---|---|
| | | MF | CO | Lard | CB |
| 1 | H | 4 | – | – | to 100% |
| 2 | D/O | 4 | 1 | – | to 100% |
| 3 | B/C/I | 5 | – | – | to 100% |
| 4 | J/Q | 5 | – | 20 | to 100% |
| 5 | S | 10 | – | – | to 100% |
| 6 | G/P | 10 | 5.5 | – | to 100% |
| 7 | M/R | 10 | 11 | – | to 100% |
| 8 | E/K/L | 12 | – | – | to 100% |
| 9 | A | 25 | – | – | to 100% |
| 10 | N | 27 | – | – | to 100% |
| 11 | F | 100 | – | – | – |
| 12 | T | – | – | – | 100 |

Notes

1.  All samples analysed as blind duplicates except B/C and K/L which were known duplicates.

2.  MF : milk fat
    CO : coconut oil
    CB : cocoa butter

The results which were obtained using the unmodified Phillips and Sanders method are given in Tables 2 and 3.

## Table 2
### Results obtained in milk fat in admixtures with cocoa butters trial

| Sample | | Laboratory | | | | |
|---|---|---|---|---|---|---|
| No. | Code | 1 | 2 | 3 | 4 | 5 |
| 1 | H | 3.96 | 3.79 | 3.97 | 3.0[c] | 4.4 |
| 2 | D | 4.17 | 3.33 | 3.98 | 3.8 | 3.8 |
|  | O | 4.00 | 3.44 | 3.86 | 4.4 | 4.4 |
| 3 | B | 5.12 | 4.85 | 5.11 | 4.2[a] | 4.8 |
|  | C | 5.36 | 5.39 | 5.13 | 3.8[a] | 5.1 |
|  | I | 4.97 | 4.31 | 4.84 | 6.1[a] | 4.8 |
| 4 | J | 5.03 | 5.10 | 4.82 | 4.6 | 5.2 |
|  | Q | 4.72 | 4.67 | 4.40 | 4.6 | 4.8 |
| 5 | S | 9.69 | 9.72 | 9.30 | 9.5 | 10.0 |
| 6 | G | 10.20 | 8.98[a] | 9.90 | 9.9 | 10.6 |
|  | P | 9.39 | 11.00[a] | 9.55 | 10.2 | 9.3 |
| 7 | M | 9.57 | 9.52 | 9.62 | 9.4 | 9.3 |
|  | R | 8.99 | 9.44 | 9.20 | 10.2 | 10.2 |
| 8 | E | 12.40 | 11.67 | 11.69 | 11.1 | 11.3 |
|  | K | 11.10 | 11.17 | 11.29 | 13.0 | 12.9 |
|  | L | 11.80 | 10.58 | 11.29 | 14.7 | 11.8 |
| 9 | A | 23.90 | 23.80 | 24.40 | 25.0 | 26.2 |
| 10 | N | 26.70 | 25.60 | 26.21 | 24.9 | 26.7 |
| 11 | F | 94.50 | 103.40 | 98.67 | 104.1 | 105.1 |
| 12 | T | 0.11 | 0 | 0.0 | 2.4 | 0.0 |

Notes

a: results rejected by Cochran's Test at $P \leqslant 0.05$ level

b: results rejected by Dixon's Test at $P \leqslant 0.05$ level

c: rejected because other Pair member rejected (see Table 4).

Table 3

Results obtained in Milk Fat in  Admixtures with
Cocoa Butters Trial

| Sample | | Laboratory | | | | |
|---|---|---|---|---|---|---|
| No | Code | 6 | 7 | 8 | 9 | 10 |
| 1 | H | 4.6 | 3.75 | 4.2 | 4.2 | 4.45 |
| 2 | D | 4.8 | 3.99 | 4.3 | 4.2 | 4.3 |
|  | O | 4.6 | 3.88 | 4.1 | 3.9 | 4.55 |
| 3 | B | 6.2$^b$ | 5.04 | 5.0 | 5.1 | 5.2 |
|  | C | 5.2$^b$ | 4.65 | 5.0 | 4.8 | 5.4 |
|  | I | 5.6 | 4.72 | 5.1 | 5.2 | 5.25 |
| 4 | J | 4.9 | 4.79 | 5.1 | 5.0 | 5.45 |
|  | Q | 5.1 | 4.93 | 5.1 | 5.1 | 5.35 |
| 5 | S | 10.1 | 9.93 | 9.7 | 9.6 | 9.9 |
| 6 | G | 10.2 | 9.69 | 9.9 | 10.1 | 10.2 |
|  | P | 9.3 | 9.75 | 9.6 | 9.9 | 10.0 |
| 7 | M | 10.2 | 9.56 | 9.9 | 10.2 | 9.85 |
|  | R | 9.7 | 9.66 | 9.7 | 9.9 | 9.75 |
| 8 | E | 14.1 | 11.47 | 11.9 | 13.8 | 11.8 |
|  | K | 11.6 | 11.33 | 11.7 | 11.9 | 11.8 |
|  | L | 12.0 | 11.74 | 11.8 | 13.0 | 11.9 |
| 9 | A | 28.7 | 24.27 | 24.7 | 26.9 | 24.5 |
| 10 | N | 26.3 | 26.11 | 26.3 | 27.6 | 26.1 |
| 11 | F | 99.5 | 96.40 | 100.6 | 113.0 | 93.8 |
| 12 | T | 0 | 0 | 0 | 1.5 | 0 |

Notes

As for Table 2

The values of the repeatabilities and reproducibilities obtained using the unmodified Phillips and Sanders method are given in Table 4; these values are as defined by the International Standards Organisation[5].

Table 4

Summary of Means, Repeatabilities and Reproducibilities
obtained in Milk Fat in  Admixtures with Cocoa Butters Trial

| Sample Combin-ation | Mean (g/100g) | Repeat-ability (r) | Reprodu-cibility (R) |
|---|---|---|---|
| a.  Uniform level analyses | | | |
| D/O ( 4%)[a] | 4.09 | 0.64 | 1.07 |
| B/C ( 5%)[a] | 5.07 | 0.60 | 0.61 |
| B/I ( 5%)[a] | 4.96 | 0.51 | 0.69 |
| C/I ( 5%)[a] | 5.00 | 0.92 | 0.92 |
| J/Q ( 5%)[a] | 4.96 | 0.66 | 0.86 |
| G/P (10%)[a] | 9.87 | 1.26 | 1.22 |
| N/R (10%)[a] | 9.69 | 0.97 | 0.97 |
| K/L (12%)[a] | 11.92 | 1.61 | 2.59 |
| K/E (12%)[a] | 11.95 | 2.70 | 2.70 |
| L/E (12%)[a] | 12.09 | 2.83 | 2.84 |
| b.  Split level analyses | | | |
| H/I (4.5%)[a] | 4.56 | 0.45 | 0.95 |
| S/C (11%)[a] | 10.96 | 1.84 | 2.17 |
| A/N (26%)[a] | 25.74 | 2.91 | 3.44 |

Notes

a:  expected value of milk fat in sample combination

The results may be considered to be satisfactory. Following
analysis of the trial results it was decided to apply the method
to milk chocolate samples on which the fat would have to be
extracted before GLC analysis could be carried out.


Milk Fat in  Milk Chocolate Trial
This trial, which took place in 1983, was arranged to test the
method used in the previous trial to fat after extraction from
milk chocolate samples by the OICC procedure[6].


18 Analysts took part in the trial ( 14 UK public analysts,
2 analysts from industrial organisations and 2 from Government
establishments).


Three samples were prepared; these were each divided into
two and given to the analyst to analyse once only as blind
duplicates.


The composition of the samples. results obtained and values
of repeatability and reproducibility calculated from these
results are given in Table 5.


CONCLUSIONS FROM TRIALS
The method used in the trials is satisfactory when applied
to pure fat samples.  It is less so when applied to milk
chocolate samples which require the extraction of fat before
the GLC analysis may be carried out.


The factor for butyric acid in milk fat is subject to natural
variability, as is the situation for all chemical indicators
of milk fat content; 3.6% was arbitarily selected as representing

Table 5

Results. Means. Repeatabilities and Reproducibilities obtained in Milk Fat in Milk Chocolate Trial

| Laboratory | Sample (g milk fat/100g sample) | | | | | |
|---|---|---|---|---|---|---|
| | A | D | B | E | C | F |
| 1 | 5.41 | 5.45 | 6.21$^a$ | 7.63$^a$ | 5.53 | 5.60 |
| 2 | 5.13 | 5.10 | 6.50 | 6.74 | 6.03 | 6.11 |
| 3 | 5.34 | 5.36 | 7.02 | 7.06 | 6.47 | 6.44 |
| 4 | 4.90 | 5.68 | 7.25 | 7.49 | 6.77 | 6.77 |
| 5 | 1.72$^b$ | 2.00$^b$ | 2.61$^b$ | 2.58$^b$ | 2.48$^b$ | 2.66$^b$ |
| 6 | 5.18 | 5.29 | 7.02 | 7.00 | 6.22 | 6.28 |
| 7 | 7.34 | 5.30 | 7.23 | 7.00 | 6.53 | 6.27 |
| 8 | 6.41 | 6.50 | 8.76$^b$ | 8.41$^b$ | 7.98 | 7.98 |
| 9 | 5.15 | 6.92 | 7.20$^a$ | 5.29$^a$ | 4.74a | 6.33$^a$ |
| 10 | 5.35 | 5.13 | 7.04 | 7.16 | 6.24 | 6.53 |
| 11 | 5.73 | 5.75 | 7.51 | 7.28 | 6.70 | 6.92 |
| 12 | 4.66 | 4.72 | 6.34 | 6.28 | 5.56 | 5.78 |
| 13 | 4.72 | 4.68 | 6.57 | 6.02 | 5.22 | 5.54 |
| 14 | 2.76$^b$ | 2.42$^b$ | 3.57$^b$ | 3.16$^b$ | 3.66$^b$ | 3.45$^b$ |
| 15 | 4.61 | 4.33 | 2.57$^a$ | 5.49$^a$ | 3.50$^a$ | 5.42$^a$ |
| 16 | 5.24 | 5.33 | 7.42 | 6.86 | 6.50 | 6.65 |
| 17 | 5.02 | 4.94 | 6.90 | 6.61 | 5.84 | 6.02 |
| 18 | 6.09 | 4.74 | 6.70 | 6.73 | 5.91 | 6.23 |
| Mean$^c$ | 5.36 | | 6.91 | | 6.31 | |
| Expected Mean$^d$ | 3.7 | | 7.5 | | 6.7 | |
| Repeatability | 1.57 | | 0.56 | | 0.40 | |
| Reproducibility | 1.93 | | 1.11 | | 1.85 | |

Notes

a: results rejected by Cochran's Test P ≤ 0.05

b: results rejected by Dixon's Test P ≤ 0.5

c: calculated using 3.6g butyric acid/100g milk fat as the factor

d: g milk fat/100g chocolate used to prepare samples

the range of possible values.  The variability of the factor
is not so great as to prevent a positive recommendation to
accept the method being made.  However, further work into other
methods for the determination of milk fat is being carried out
and it is possible that a method based on triglyceride analysis
of the extracted fats will be proposed   Initial results suggest
that milk fat contents calculated using the results of such analyses
appear to exhibit less natural variability than those obtained
using butyric acid as the indicator of milk fat concentration.

REFERENCES

1.    Council Directive 73/241/EEC, O. J. No. L 228, P.23,
      16.8.1973.

2.    "The Cocoa and Chocolate Products Regulations 1976",
      SI 1976, No. 541.

3.    Office  International du Cacao et du Chocolate (OICC),
      Analytical Method 8i/1960.

4.    A. R. Phillips and B. J. Sanders, J. Assoc. Publ. Analysts,
      1968, 6, 89.

5.    International Organisation for Standardization.
      "Precision of Test Methods" ISO/DIS 5725:1977.

6.    Office International du Cacao et du Chocolate (OICC).
      Analytical method 8a/1972.

## Appendix I

GLC conditions used in the Phillips and Sanders procedure

1.  GLC Column : 5 ft x ¼ in. o.d. (glass)

2.  GLC Column Packing:

> 5 per cent Carbowax 20 M + 0.5 per cent
> terephthalic acid on 100–120 mesh,
> acid-washed Supasorb.
>
> (Prepare as follows:- reflux ml of ethanol
> with 4 g of Carbowax 20 M and 0.4 g
> terephthalic acid until dissolved. Add 20 g
> of Supasorb and boil until reflux to remove
> air. Filter rapidly at the pump
> (approximately 25 ml of the solution is
> retained by the Supasorb) and dry the residue
> under vacuum. After packing the columns,
> purge with nitrogen at $220^{\circ}C$ for 24 hours).

3.  GLC Operating temperature: $125^{\circ}C$.

4.  GLC Detector: f.i.d.

# Determination of Tartrazine in Rice Milk Desserts by HPLC Following Ion-pair Extraction with Tri-n-octylamine. Evaluation by Collaborative Study

By L. Dryon, D. L. Massart, and M. Puttemans*

PHARMACEUTICAL INSTITUTE, VRIJE UNIVERSITEIT BRUSSEL, LAARBEEKLAAN 103,
B-1090 BRUSSELS, BELGIUM

Toxicological investigation of food colors is underway in many countries[1,2].These toxicological data result in repeated revisions of the number of permitted food dyes.This number decreases faster in countries having a more restrictive legislation.Systematic studies on chronic toxicity have led pharmacologists and nutritionists to define acceptable daily intakes(ADI)for each of the permitted colorants.This evolution has recently led to more restrictive legislations which impose a maximum allowed concentration of a dye in a given foodstuff,as is the case in Belgium[3].For example, the limit of tartrazine in rice milk desserts is 100 mg/kg.Consequently,there is a need for an accurate and precise analysis scheme consisting of an extraction,or sample clean-up,followed by a selective separation and identification+dosage technique.

Because of their sulfonic and,in some cases,carboxylic acid functions,food dyes are hydrophilic and are present as anions over a large pH range.Such substances are difficult to extract by suppression of ionization but can be well extracted by ion-pair formation.In ion-pair extraction,an ionized solute is extracted into an organic phase after addition of a suitable counterion.The counterion,a hydrophobic ion of opposite charge,forms a hydrophobic complex with the solute which is then extracted to a higher degree than the solute itself by the organic phase.The organic phase is usually chloroform because of its ability to solvate the ion-pair formed[4].

The equilibria which will compete in the extraction of the acid HX can be described as follows[5]:

1.<u>Dissociation of the acid HX in the aqueous(subscript a)phase:</u>

$$HX_a \rightleftharpoons H_a^+ + X_a^-$$

2.<u>Distribution of the acid HX between the aqueous and the organic (subscript o) phase:</u>

$$HX_a \rightleftharpoons HX_o$$

3.<u>Extraction of the anion $X^-$ by ion-pair formation with the counterion $Q^+$:</u>

$$X_a^- + Q^+ \rightleftharpoons QX_o$$

The counterion may be added to the aqueous as well as to the organic phase!

The role of the pH of the extraction medium is double;both the solute and the counterion must be ionized.Anionic dyes can be extracted by quaternary ammonium compounds or amines[5].

In some cases it may be necessary to extract the analyte back to an aqueous phase;this can be achieved by the following displacement reaction:[6,7]

4.<u>Displacement of $X^-$ by an other anion $Y^-$:</u>

$$QX_o + Y_a^- \rightleftharpoons QY_o + X_a^-$$

Several displacing ions have been used,<u>e.g.</u>chloride,bromide,iodide,nitrate and perchlorate.Highest recoveries were obtained with perchlorate[6].

\*\*\*\*\*\*\*\*\*\*\*\*\*\*\*\*\*\*\*\*\*\*\*\*\*\*\*\*\*\*\*\*\*

Optimum recoveries of dyes were obtained when the following parameters were respected:

  a pH of 5.5

  extraction with tri-n-octylamine

  a counterion concentration of 0.1 M

  chloroform as the solvent

  perchlorate as the displacing ion(0.1 M)

\*\*\*\*\*\*\*\*\*\*\*\*\*\*\*\*\*\*\*\*\*\*\*\*\*\*\*\*\*\*\*\*\*

The method was already employed for alcoholic beverages[6].When more complex foodstuffs have to be analyzed the dyes have to be liberated first from the food matrix.In the case of rice milk desserts this can be achieved by an elution with methanol-ammonia mixtures[8].This elution was performed as follows:2gr.of dried rice milk(dried by lyophilization)is transferred to a glass chromatography tube;the dyes are eluted with varying amounts of methanol-

ammonia mixtures.The determination is then carried out as described in the appendix.The influence of the diameter of the
chromatography tube,the eluent composition and volume are given
in Tables 1 and 2:[8]

Table 1:Recovery of tartrazine in rice milk(50 mg/kg)
        using 1 and 2 cm columns,as a function of the
        eluent volume(methanol-ammonia(95:5))

|                  | Recovery % | |
| Eluent,ml | 1 cm | 2 cm |
| --- | --- | --- |
| 25  | 71.9  | –    |
| 50  | 73.2  | –    |
| 100 | 100.3 | 24.0 |
| 200 | –     | 26.2 |
| 300 | –     | 30.3 |

Table 2:Recovery of tartrazine from rice milk(50 mg/kg)
        as a function of composition and volume of the
        eluent(methanol-ammonia)

| Eluent | Recovery % | | |
| (% methanol) | 25 ml | 50 ml | 100 ml |
| --- | --- | --- | --- |
| 100 | $25.3^{\pm}1.4$ | $58.6^{\pm}7.9$ | $100.0^{\pm}3.4$ |
| 95  | $71.9^{\pm}0.5$ | $73.2^{\pm}0.2$ | $100.3^{\pm}0.8$ |
| 90  | $79.1^{\pm}1.6$ | $93.7^{\pm}5.1$ | $100.9^{\pm}1.6$ |
| 85  | $67.0^{\pm}4.5$ | $82.3^{\pm}4.1$ | $86.0^{\pm}0.7$ |
| 80  | $82.0^{\pm}2.0$ | $89.9^{\pm}3.8$ | $87.3^{\pm}1.8$ |

From the above given data the following conclusions were
drawn[8]:
   -a 1 cm chromatography tube is preferable
   -at least 100 ml of eluent is required
   -a methanol-ammonia ratio of 95:5 gives excellent recoveries

******************************

The separation technique used for identification and determination purposes is reversed phase High Performance Liquid Chro-
matography.As stated before,dyes are hydrophilic substances.Such
solutes have short retention times in a reversed phase HPLC system.
A counterion,*i.e.*tetrabutylammonium,was therefore added to the mobile phase in order to increase the retention of the dyes[6,8]

The analysis scheme used(fully detailed in the Appendix!)is
specific for tartrazine since the combination of the TnOA extraction followed by the perchlorate back-extraction extracts only
anionic dyes and,furthermore,tartrazine is resolved from other
dyes by the HPLC system[8]

********************************

Based on all these observations a collaborative study was
started in agreement with AOAC's General Referee for Synthetic
Dyes.Potential participants to this study were invited to state
their willingness to collaborate or not.9 laboratories responded
in a positive way.To each collaborator were sent:
   -10 samples,divided as follows
                  1 blank
                  4 spiked samples(3 concentration levels)
                  5 commercial samples(4 different)
   -the ion-pairing reagent(Tri-n-Octylamine)
   -the tartrazine standard
   -the analysis procedure
   -a list of reagents and equipment needed


The collaborators were spread as follows:
                  -3 in Belgium
                  -1 in Switzerland
                  -1 in England
                  -1 in Canada
                  -3 in the United States


In Appendix is given the full text that was sent to the col-
laborators.At the present moment(December 1983)it is impossible
to present data about the evaluation of the method since only 4
collaborators transmitted their results.Full results will however
be given at the Seminar.

Literature List

[1]K.Khera and I.Munro,<u>CRC Critical Rev.Toxicol.</u>,1979,81

[2]K.Venkataraman,"Analytical Chemistry of Synthetic Dyes",Wiley, New York,1977

[3]Belgian Food Legislation,KB 27/7/1978

[4]G.Schill,"Separation Methods for Drugs and Related Organic Compounds",Apotekarsocieteten,Stockholm,1978

[5]M.Puttemans,L.Dryon and D.L.Massart,<u>Anal.Chim.Acta</u>,1980,<u>113</u>,307

[6]M.Puttemans,L.Dryon and D.L.Massart,<u>J.Assoc.Off.Anal.Chem.</u>,1982, <u>65</u>,737

[7]M.Puttemans,L.Dryon and D.L.Massart,<u>J.Assoc.Off.Anal.Chem.</u>,1982, <u>65</u>,730

[8]M.Puttemans,L.Dryon and D.L.Massart,<u>J.Assoc.Off.Anal.Chem.</u>,1983, <u>66</u>,720

APPENDIX

(This appendix contains the text that was sent to the collaborators of this study)

Collaborative study:

"Determination of tartrazine in rice milk by HPLC following ion-pair extraction with Tri-n-Octylamine."

The following text contains a list of reagents and apparatus that are necessary for this analysis.
Please check if you have all of them,if not let us know.

Samples have been lyophilized previously and should be stored (in their original container) preferably in a dessicator in order to prevent water uptake.

This is a study of the method,not of the laboratory.
The method must be followed as closely as practicable,and any deviations,no matter how trivial they may seem,must be noted on the report form!

## Apparatus:

-Glass chromatography tubes with an internal diameter of 1 cm
   (length _ca_.40 cm)and equipped with a teflon tap or another
   device to control the elution rate,the column is also
   equipped with a fritted disk pore size _ca_.0.05 mm.
-Stoppered glass centrifuge tubes with a content of 30 ml and
   10 ml
-Mechanical shaking apparatus:operates at room temperature and
   at a rate of 60 shakings/minute
-Centrifuge capable of centrifuging at 2000 rpm.
-pH meter
-HPLC system equipped with:
   -loop injector(0.1 ml)
   -UV detector 254 nm or variable wavelength detector operating
   at _ca_.420 nm.
   -25 or 30 cm column packed with 10 microm.octadecylsilica
   (i.d.4mm) (suggested:RP-18 Lichrosorb,Merck)
   -flow rate:1 ml/min,roomtemperature
-Integrator

## Reagents:

-Methanol,ammonia 25%,sodiumphosphates,chloroform,sodiumperchlo-
   rate,phosphoric acid and sodiumhydroxide are analytical grade
   reagents.
-Tetrabutylammonium:several possibilities exist
                  -phosphate 0.5 M aqueous solution(Altex)
                  -hydroxide 25% in methanol(Fluka)
                  -PIC reagent(Waters)
-Tartrazine standard:will be supplied
-Tri-n-Octylamine:will be supplied
 (is hygroscopic,store in brown bottle in a dessicator)
 a 0.1 M solution in chloroform is prepared
-Composition of buffers:

|          | g $NaH_2PO_4.H_2O$ | g $Na_2HPO_4.2H_2O$ |
|----------|--------------------|---------------------|
| pH=5.50  | 24.65              | 1.26                |
| pH=7.00  | 5.77               | 9.38                |

dissolve in bidistilled water and dilute to 1.8 l;if necessary
the pH is adjusted to 5.50 $\pm$ 0.05 and 7.00 $\pm$ 0.05 with 0.1 M
$H_3PO_4$ or NaOH,the buffer is diluted to 2 l.
-Mobile phase:
    methanol - phosphate buffer pH=7.00 mixture which contains
    0.005 M tetrabutylammonium
The suggested composition for the RP-18 Lichrosorb(Merck)column
is methanol-phosphate buffer (40 + 60).


Preparation of mobile phase:
———————————————

Mix 40 volume parts of methanol with 60 volume parts of buffer.
The eluent is prepared by diluting a sufficient amount(or volume)
of a tetrabutylammonium solution to produce a 0.005 M solution.
Eluents are always filtered through a 0.45 μm filter(Millipore).

The composition of the mobile phase,i.e.the methanol content,may
be adjusted in such a way that the retention time of tartrazine
will be ca.5 min.(flow rate 1 ml/min).
The system should also separate the following mixture to the
baseline:indigotine(FD&C Blue N°2),amaranth(FD&C Red N°2) and
tartrazine.A chromatogram of this separation should be supplied
with the final report.


Procedure:
—————————

The rice milk was initially dried by lyophilization and was
homogenized by rubbing it to powder in a mortar.Each sample
is analyzed once.

- 2 g of rice milk powder is transferred to the glass chromatography tube.

-The dye is eluted with 10 fractions of 15 ml of a methanol/
  ammonia (95:5) mixture.The elution rate is adjusted to 0.5 -
  1 ml/min.
The top of the column may never come dry except with the last
fraction of eluent.

-The eluate is collected in a 200 ml erlenmeyer and is evaporated on an electrically heated plate (temperature ca.40°C)
placed in a fume chamber.
At the end of this evaporation the residue may not carbonize!

-The residue is redissolved in 20 ml of phosphate buffer pH=5.50
and is transferred to a 30 ml centrifuge tube which contains 5
ml of a 0.1 M Tri-n-Octylamine solution in chloroform.

-The tube is shaken for 30 minutes;the tube is centrifuged at
2000 rpm for 15 minutes.

-The upper layer is discarded.

- 3 ml of the chloroform layer are transferred to a 10 ml centrifuge tube,3 ml of 0.1 M sodiumperchlorate solution is added,
the tube is shaken for 30 minutes and centrifuged at 2000 rpm
for 15 minutes.

- 0.1 ml of the upper layer is injected three times in the HPLC
system;peak areas are calculated and averaged.

-A calibration curve is constructed by plotting the area of the
tartrazine peak versus its concentration;tartrazine standards
are prepared in bidistilled water:10,20,30,40 and 50 ppm;each
standard is also injected three times.
The average peak areas are used in the construction of the
calibration curve.
The correlation coefficient,the slope and the intercept of
the curve with the Y axis are calculated.The amount of tartrazine in the samples is calculated by linear regression analysis
(least squares).

Report:

Supply all the results obtained:recorder plots

graphs of the calibration curve

integration results

The report should be given as follows:


Name of the laboratory:Samples A (B,C,...)

| Nr. of sample | Weight(grams) | Area | Conc. extr.mg/l | Amount in rice milk mg/kg |
|---|---|---|---|---|
| 1 | | | | |
| 2 | | | | |


Amount of tartrazine in dried rice milk= $\dfrac{\text{conc.extract(mg/l)} \times 5}{\text{weight(grams)}}$

(mg/kg)

# Assessment and Optimization of Indirect Instrumental Methods for Testing Major Constituents in Milk and Dairy Products

By R. Grappin
I.N.R.A., DAIRY EXPERIMENTAL STATION, 39800 POLIGNY, FRANCE

Like many other fields, dairy analyses have changed dramatically during the past 10 or 15 years. The number and nature of tests performed regularly have increased, and now in many countries dairy laboratories are using instrumental methods instead of the classical manual techniques for testing major components in milk and milk products. By major components, it is usually meant the components which are quantitatively important like water, fat, protein and lactose. However, the nature of the components which are the most commonly tested varies according to their variability and economic values.

Taking advantage of their high rate of analysis and automaticity, instrumental methods are used mainly for mass testing, within the framework of milk payment and herd improvement programmes, and for quality control and on-line process control in the dairy plants.

In Table 1 are listed the principal instruments currently used for the measurement of fat, protein, lactose and total solids in milk and dairy products. Most of these instruments, except Near Infra-Red Reflectance (N.I.R.) and Microwave ovens, were primarily designed for milk analysis and are not usually directly applicable to the analysis of dairy products, especially solid foods. Now, for mass testing, Infra-Red (I.R.) instruments prevail over the other techniques, which tend to be used either as secondary reference methods for instrument calibration or as routine methods of analysis in the dairy industry.

Looking at the high speed and degree of accuracy, versatility and automaticity of many of the instruments currently used by laboratories, one would deem that the best possible in instrumentation has been achieved. However, for the laboratory manager, the scientist or the regulatory agency who is directly involved in dairy analysis, many problems remain and improvements are desirable. Cardone (2) has recently mentioned that the operation of method verification invokes both the process of method validation : how accurate is it ? and the method evaluation programme : how good is the method in the end user's hands ? In this presentation, I will stress some statistical aspects of analytical performances assessment, with special emphasis to systematic errors, give examples of the optimization process of instrumental method, and point out some practical

problems encountered by dairy laboratories for the control and the calibration
of instruments. Finally, I will evoke some recent progress in instrumentation
and a possible future for the analysis of major components.

Table 1 : Principal indirect instrumental methods used for testing major components in milk and milk products
(1) list non-exhaustive

| Methods | : | Instruments available (1) | : | Component measurement |
|---|---|---|---|---|
| Dye-binding | : | Pro-Milk | : | Protein |
|  | : | Udy Analyser | : |  |
| Turbidimetry | : | Milko-Tester | : | Fat |
| Infra-Red Spectroscopy | : | I.R.M.A. Milko-Scan Multispec | : | Fat, protein, lactose, water |
| Near Infra-Red Reflectance Spectroscopy | : | Infra-Alyser | : | Fat, protein, lactose, water and others |
| Microwave Drying | : | Appolo TMS Checker | : | water |

## 1. Assessment of Analytical Performances

Assessment of analytical performances is the final step of a method development programme, taking place after all sources of error have been identified
and if possible eliminated or reduced. The relative importance of the performance characteristics given in Table 2 varies with the method itself, the purpose
of the analysis and the concentration of the analyte. For instance, the limit
of detection which is a critical parameter in trace analysis, where the component concentration is in part per million or lower, has never been considered
as a very important attribute in the analysis of constituents which are in the
range of concentration of the percent. Similarly, a relatively good precision
can be expected in the analysis of these major constituents, because of the inverse relationship existing between precision and concentration. Horwitz et al.
(7) showed that for every 100-fold decrease in concentration of the analyte

the coefficient of variation of precision increases by two-fold. Moreover, precision of automated methods is theoretically better than the repeatability of the corresponding manual techniques because all the steps of the analytical process are automatically controlled by mechanical or electronic devices, thus reducing considerably the risk of random errors. In fact, for all the instruments listed in Table 1, a coefficient of variation of repeatability of 1 % or lower is generally obtained.

Table 2 : Performance characteristics, or attributes, of analytical methods (mainly from ISO definitions)

| | TERMINOLOGY | : | FUNCTIONAL DEFINITION |
|---|---|---|---|
| ACCURACY | : PRECISION (Repeatability | : | within-laboratory variability |
| | :(Random error)(Reproducibility: | | between-laboratory variability |
| | ACCURACY OF THE MEAN (Systematic error) | . : | Mean difference from the true value |
| | SENSITIVITY | : . | Ratio measured value/concentration |
| | :LIMIT OF DETECTION | : | Smallest concentration detected |

Contrary to the limit of detection and precision experience has shown that the systematic part of errors or bias is the main source of analytical errors of instrumental methods. Discussion will be limited to some statistical and practical aspects of the evaluation of these systematic errors.

1.2. Evaluation of systematic errors. Instrumental methods are usually considered indirect methods because they do not measure directly the component they are intended to measure but instead measure one or more quantities or properties which are functionally linked to that component. For instance, the Milko-Tester used for milk fat testing does not measure directly the fat content, but measures a related property, the turbidity produced by milk fat globules. Quite often the measured quantities or properties are not exactly proportional to the component concentration, introducing therefore systematic errors in the final test results. Indirect instrumental methods of analysis can yield different kinds of systematic errors :

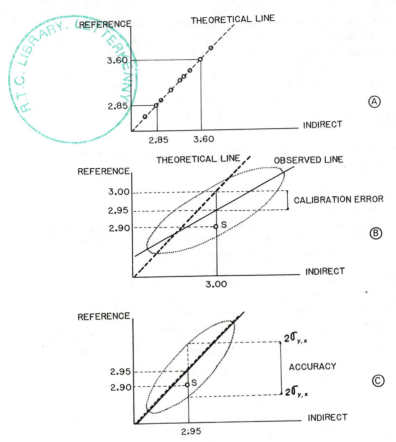

Figure 1 : Calibration error and accuracy of indirect method

Figure 1.A shows the ideal situation for an indirect method, where the plot of
the reference vs. the indirect results gives a straight line passing through
the origin with a slope of 1.000. All the data points, which are the average
of replicates, are located on the line, and at each level, the instrumental values equal the reference values.

The ellipse in **Figure 1.**B represents a population of samples, analyzed by the
reference method and by the instrument, which is, this time, not correctly calibrated. The observed line  is obtained from the regression of reference vs.
indirect results.

For a given sample (S) the difference between the reference value and the indirect value (3.00 - 2.90 = 0.10 %) can be split into two components or two distances : the distance from the point (S) to the actual regression and the distance from the regression to the theoretical line; this latter distance represents the calibration error at 3.00 % level.

Figure 1.C shows the same situation, but the instrument is now perfectly calibrated. The observed line and the theoretical line are now coincident, eliminating the calibration error, not only for the sample (S) but for all the samples population.

The indirect value, which is shifted from 3.00 to 2.95 % by adjusting the calibration, still does not match the reference value of 2.90 %. The difference between the two methods (0.05 %) is closely associated with the particular physico-chemical characteristics of the sample (S).

From the property of the standard deviation from the regression ($s_{y,x}$), 95 % of the population of samples lie within the tolerance interval of $\pm 2\ s_{y;x}$. This value represents an important attribute of the method : the accuracy. It gives a correct figure of the predicting value of the indirect method and is independent of the exactness of calibration.

Now in many experimental works only the coefficient of correlation is used to indicate how close the reference and the indirect methods are. It has been already stressed (4) that the coefficient of correlation can be misleading because its value depends as much on the range of sample population ($s_y^2$) as it does on the deviation from the regression ($s_{y,x}^2$) From the formula : $r = (\ 1 - s_{y,x}^2\ /s_y^2)^{1/2}$ , we can see that it is possible to get an impressive value of r, only by expanding the range of the variation of the variable y (large $s_y^2$ value), with samples having unusually high and low concentrations, whereas the s.d. of estimate, which represents the real accuracy, remains constant.

When the objective of a comparison between instrumental and reference values is a proficiency evaluation of a laboratory or instrument, the mean and the s.d. of the algebraic differences between the two methods should be used because it includes both aspects of the systematic errors: calibration and accuracy. When an instrument is correctly calibrated, the mean difference is zero and the s.d. of the differences is approximatively equal to the s.d. of the accuracy ($s_{x-y}^2 \approx s_{x,y}^2$), indicating a correct slope (**Figure 2**).

From a survey of the literature it was found that the relative standard deviation of accuracy of the various instruments used for measuring fat, protein, lactose and total solids in milk and dairy products varies, according the product and the method, from 0.5 % up to 2 %.

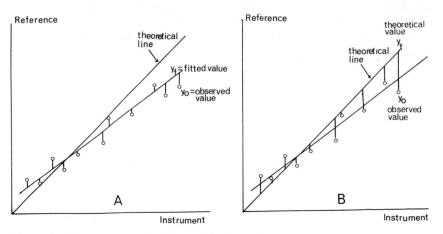

Figure 2 : diagramme showing the real instrument accuracy represented by the va--
riance of residuals $y_o - y_f$ (A), and the actual instrument accuracy represented
by the variance of the differences $y_o - y_t$ (B).


In order to achieve the most accurate calibration, it is essential to detect
which milk component or property, other than those measured by the instrument,
may interfere on the instrument response, and then to determine which factors
are responsible for their variation. This experimental work represents an important part of the optimization procedure of the method.


## 2. Optimization

        The AOAC Comittee on interlaboratory studies has recently stressed (8)
that the procedure to validate the performance of a method shoud be conducted
only on methods which have been optimized.
According to Dols and Armbrecht (3), method optimizatation is the process of
finding the optimum operating conditions of a method, and this process involves
three stages : firstly obtaining a response , secondly improving the response and
finally understanding the response, and unfortunately, they added ,too often only
the first stage is completed.
Instruments must be tested in the most variable conditions to get a better detection of all sources of errors : instrumental and environmental factors (temperature, humidity, homogenization, etc.), sample matrix and ageing effects.

Instrumental and environmental factors are now relatively well known and can be controlled through repeatability and stability tests. Relying upon the princi- ple of each method it should be possible to determine the nature of the inter- ferences or matrix effect. Considering for instance the I.R measurement of milk fat at 5.73 µm based upon the absorption of energy by the ester linkages of the triglycerides, one can predict the accuracy of the method to be influen- ced by factors like the degree of lipolysis or the average triglycerides mole- cular weight. Several experimental works have confirmed and quantified the in- fluence of these two factors. An increase of one unit of the lipolysis index (B.D.I. Acid Degree Value) lowers I.R fat test results at 5.73 µm, by 0.0224 % ( Figure 3).

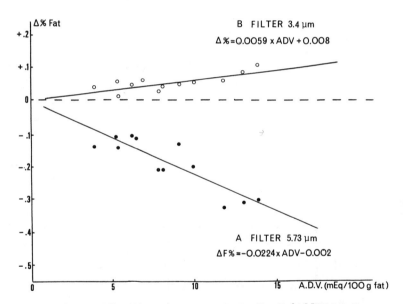

Influence of Natural Lipase on Infra Red Fat Test Results (at 3.5% fat level)

- Figure 3 : Influence of milk lipolysis on Infra-Red fat test
  - Δ % : difference of test results between lipolyzed and reference (fresh) sam- ples

Considering milk fat refractive index proportional to the average fatty acids molecular weight (9), **Figure 4** shows that an increase of $10^{-4}$ unit of the refractive index decreases I.R. signal by 0.20 %, in relative value.

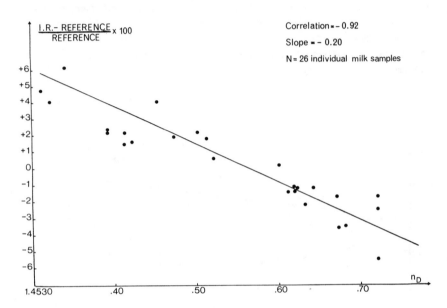

Figure 4 : Influence of milk fat refractive index on Infra-Red (5.73 µ m) fat determination
- 100 (I.R.-Reference) / Reference : difference on relative value (%) between Infra-Red and Röse-Gottlieb fat test results
- $n_D$ : refractive index

When the study of the influence of the various physico-chemical characteristics of the product is completed, it is essential to determine the mangnitude of their natural variation and the consequences on the accuracy and calibration of the instrument. This second step has been too often overlooked, or it was assumed that the natural variations of milk characteristics were not sufficient to alter significantly the accuracy of the analyses. Now, most of the biological factors, like season, feeding. stage of lactation, species have been studied (11) and we know what factors can be conteracted by an appropriate ca-

libration. Figure 6 gives an example of the influence of the season (a mixed
influence of feeding and stage of lactation) on the accuracy of fat test of
ewe's milk.


## 3. Practical applications : correction of random and systematic errors

Analytical errors are usually established by comparing instrumental and
reference test values obtained either from a single sample or from the mean of
a population of samples. When the difference between the two values is beyond the
accepted accuracy, it is usually concluded that the instrument needs to be re-
calibrated. Before drawing such a conclusion it is essential to determine whether the difference originates from an instability of the instrument response, or
from a change in the calibration function, i.e. a change in the relationship
between the instrument signal and the reference value. Between these two sources of errors, which are too often mistaken, the checking procedure is different
and the action to be taken is different.

## 3.1. Stability of Instrument.
An instrument is stable when the response produ-
ced by a given impulse, represented here by the analysis of a control milk, remains over a period of time within the confidence limits of the short-term reproducibility of the instrument. A good stability indicates that the instrument
is working properly and that the instrumental and environmental conditions which
may influence the instrument signal are constant.
To monitor the instrument stability over a working day, a control chart based
upon the principles used routinely by the industry to monitor output of production lines is a very useful tool. Figure 5 shows how a control chart can be done
and how it works.
When a drift of the instrument is noticed, i.e. when the cumulative mean of the
control milk values fall outside the same confidence belt for two consecutive
samples, the operator should check the instrument functions : optical zero, temperature, homogenization, air humidity, etc., and then take the appropriate
steps. It is of course essential that no spoilage of the control milk samples
occur during the normal period of use. Normally, the instrument calibration
functions should not be changed when an instrument drift is observed.

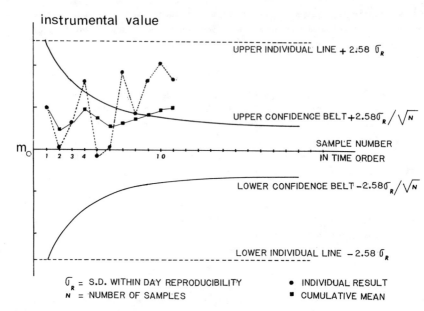

Figure 5 : Control chart for instrumental methods
- $m_0$ : standard or target value of the control milk
- upper and lower individual lines : 1 % probability of the two-sided tolerance
interval of individual test results
- upper and lower confidence belts : 1 % probability of the two-sided confidence
interval of the cumulative mean


3.2. Calibration. Instrument calibration has always been a major worry for many
users, probably because this question implies several expects. Let us take the
example of Infra-Red milk analysis. At selected wavelengths, pure solutions of
fat, protein or lactose produce primary signals which are proportional to their
concentrations. In the same conditions, milk fat, protein and lactose which are
in an heterogeneous multicomponent system   produce primary signals which are no
longer proportional to their true concentrations, essentially because of the
interference of other components.
To conteract the influence of the sample matrix  the instrument is then calibrated. By definition, calibration concerns the adjustment of the instrument signal
so that at each level of the component concentration the mean of individual
tests results given by the instrument is closely approximative to the true value of the component concentration (see **Figure** 1 C). With modern I.R. instruments, calibration is achieved in two steps corresponding to two different kinds

of interference :

1/ the interference of other measured components, e.g. lactose and protein at
the fat wavelength, which is compensated by built-in correction factors. These
factors are calculated by the analysis of manipulated milk samples obtained by
addition or removal of one or two components. When known, these correction fac
tors should not be changed.

2/ The interference of known or unknown milk components or properties, like fatty acid composition, protein composition, salt content, component hydrolysis,
etc., which are not measured by the instrument, and therefore cannot be eliminated by built-in correction factors.
The purpose of the calibration regularly performed by laboratories is precisely
to reduce or eliminate the influence of these non-measured components or factors. Obviously, such calibration will give solely an average correction  representative of the average composition of the milk samples population, but will
not correct the variability in composition of individual samples composing the
milk samples population (see **Figure** 1 C). In practice  comparison between instrument and reference results should be done on a regular basis, e.g. every
week or two weeks, with bulk milk samples over a large range of component concentration. The slope of the regression equation, and the mean of the instrumental values, must be compared to the theoretical values, and, if necessary, the
calibration is adjusted with the appropriate instrumental means.


4. Recent progress and future

A method verification programme is a dynamic process, which means that a
constant feedback from the user to the research laboratory and to the manufacturer is necessary. This dynamic process has lead directly or indirectly to real
progress in instrumentation and quality control program in dairy laboratories.
Here are two examples. To reduce the influence of fatty acid composition on I.R
fat testing, a new wavelength, called B filter, measuring $CH_2$ groups at $3.48\ \mu m$
was introduced recently. Figures 6 and 7 show clearly that using the B filter
instead of the measurement at $5.73\ \mu m$  lowers considerably the influence of two
factors chosen as examples, season and stage of lactation, improving then the
accuracy of the method.

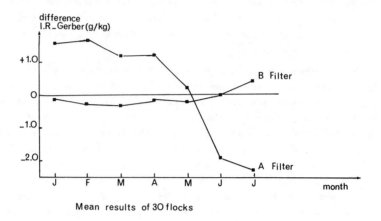

Figure 6 : Influence of season on Infra-Red fat test of ewe's milk

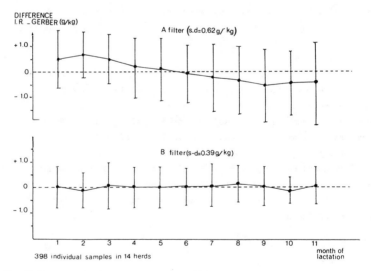

Figure 7 : Influence of the stage of lactation on Infra-Red fat test
- s.d. : residual standard deviation from ANOVA

Not directly associated with instrumental methods, computers are now used widely for data capture and processing, but they can also be used with profit for calibration instead of the electronic or mechanical devices, and for control of instrument stability. This is now done with success by many routine laboratories, not only for mass testing but for the analysis of dairy products by I.R. methods.

In many analytical areas, improvement of the overall accuracy was achieved by a considerable increase of the cost of the analysis. Before developing new equipment, it is essential that the economist, the legislator or the product manager provides the analyst a figure below which the error has no economic, legal, or technical value. However, considering for instance the accuracy and the prices and rates of analysis of the first IRMA and of the latest I.R equipments, we observe that we have now more accurate analytical results at a much lower cost per sample. However, seeking constantly a higher degree of accuracy, one should keep in mind that in the final figure, giving for instance the average composition of milk supplies, the analytical error is always small compared to the sampling errors. This statement holds only when we consider random analytical errors which are of a less pratical and economic importance than systematic errors. Reduction of systematic errors, especially those which cannot be eliminated by calibration, remains a major goal. A method with no significant matrix effect has the advantage 1/ to give results with no or low bias and 2/ to enable a calibration with standard reference materials, reducing greatly the risks of systematic differences between laboratories. Such a centralized procedure has been used for about 10 years in France for the calibration in true protein of the dye-binding method, improving considerably the reproducibility of the method. In a recent survey over 54 dairy laboratories, a between-laboratory s.d. of 0.024 % was obtained.

Can we imagine that other major compounds than water, fat, total nitrogen and lactose will be analysed regularly in the near future ? Since the development of milk payment on a protein basis, cheesemakers are more and more concerned by the cheese yielding capacity of raw milk. Is routine testing of casein or rennet coagulable protein worth doing ? Speaking only about methodology it was shown that casein or rennet casein can be tested rapidly with a great accuracy by dye-binding or I.R. methods (6,10). For instance in a recent study (6), the s.d. of accuracy of rennet coagulable protein of individual goat milk measured by a Milko-Scan was 0.026 % vs. 0.036 % for true protein and 0 066 % for total nitrogeneous matter.

The current trends show that I.R. and N.I.R. methods are more and more used by dairy laboratories. From recent works (1.5) it seems that NIR is a promising technique for the analysis of many kinds of dairy products. It does not mean

that the other methods listed in Table 1 are no longer of interest. Many small
or large dairy plants have purchased recently manual amido-black instruments ·or
microwave ovens. These techniques are reliable and accurate and  for the determinations of single component are much easier to use than the I.R. instruments

    Other instrumental methods, like G.C. or HPLC can be used for lactose and
protein determination, but it is unlikely that they would replace in the near
future I.R. methods for routine or mass testing in the dairy laboratories.

References

(1) R.S. Baer, J.F. Frank and M. Loewenstein.1983. J. Assoc. Off. Anal. Chem.,
66, 858-863

(2) M.J. Cardone, 1983. J. Assoc. Off. Anal. Chem., 66, 1257-1282

(3) T.J. Dols and B.H. Armbrecht, 1982. AOAC Working Document K

(4) T. Fearn and B.G. Osborne, 1982. FMBRA Bulletin n° 5, 222-232

(5) J.F. Frank and G.S. Birth, 1982. J. Dairy Sci., 65, 1110-1116

(6) R. Grappin and R. Jeunet, 1979. Le Lait, 59, 345-360

(7) W. Horwitz, S.L. Kamp and K.W. Boyer, 1980. J. Assoc. Off Anal. Chem.,
63, 1344-1354

(8) W. Horwitz, 1983. J. Assoc. Off. Anal. Chem., 66, 456-466

(9) H.F. Kerkhof Mogot, J. Koops, R. Neeter, K.J Slangen, H. Van Hemert,
O. Kooyman and H. Wooldrik, 1982. Neth. Milk Dairy J. , 36  115-134

(10) T.C.A. Mc Gann, A. Mathiassen and J.A. O'Connell, 1974. Lab. Practice , 21,
628-631, 650

(11) L.O. Sjaunja, 1982. Swedish University of Agricultural Sciences, Report 56

# Near Infra Red Reflectance Analysis Applied to Dairy Products

By R. W. V. Weaver

ST. IVEL TECHNICAL CENTRE, ABBEY HOUSE, CHURCH STREET, BRADFORD-ON-AVON,
WILTSHIRE BA15 IDH, U.K.

Abstract

Near infra red reflectance analysis calibrations have been
derived for the Technicon InfraAlyser for the determination
of moisture and fat in milk and milk powder, for moisture
in soft cheese and butter and for fat in cream.

The method provided a rapid means of measuring these
constituents for process control during the manufacture
of the above products.

Introduction

The composition of dairy products may be controlled by
law or by the customer's specification for parameters
such as moisture or fat content.  The economics of
manufacture demand that final products are as near the
target values as possible.
Official reference procedures take too long to have any
value for most quality or process control purposes.
Traditionally therefore within the industry a wide
range of empirical tests have been built up to measure
these parameters rapidly.

Modern Dairies now produce large quantities by continuous
means.  This has lead to demands by managements for
"instant" or "near instant" measurements so that production to an incorrect composition can be corrected immediately thus avoiding large volumes of unsatisfactory
product being produced.  Near infra red reflectance

analysis using Technicon 400D and 300 InfraAlysers has
proved an accurate means of providing fast analytical
results.

A further advantage of the technique is that the very
speed of analysis often enables more tests to be carried
out thus providing tighter control than might otherwise
be possible.

However as with all instrumental methods the results
need to be monitored by the use of standard samples.

Operating Procedure

The Technicon InfraAlyser comprises a near infra red
source and filter wheel to provide  10 or 20 individual
wavelength filters.  Infra red light passes <u>via</u> one of
the filters onto the sample, reflected light is collected
<u>via</u> a gold sphere and the intensity is measured. The
filter wheel is then moved automatically to the next
relevant filter position and the cycle repeated until
necessary positions have been read.  The InfraAlyser
then calculates the result:-

% Constituent $-$ $F_1$ Log R, + $F_2$ Log $R_2$ .......+$F_0$

Where F   is a constant from regression
      R   is a logarithum of reflectance

The number of wavelengths to be used and the $F_0$
values are determined at the time of calibration.
The F values are stored in the instrument after
calibration.

Solid samples are presented to the InfraAlyser in a
closed or open cup.  The cup is placed in a sample drawer:
closing the drawer automatically starts the analysis.
It is reflectance of the light from the surface
of the sample which is measured.  Liquids can be
analysed on the 400 Dairy (D) system; in this case

they are pumped <u>via</u> a homogeniser to a temperature-
controlled liquid cell. The reflected light measured
in this case comes both from the surface of the liquid
and from the rear of the cell. In the latter case it
has passed through the thin film of liquid contained
in the cell. In use the liquid cell fits into the same
position in the instrument as the solid sample cup.

Calibration is achieved by taking a series of samples
which have been analysed by reference procedures for
parameter or parameters of interest. These samples are
presented to the instrument and the logs of their
reflectances at each wavelength are noted. It is
important that the samples going into the calibration
exhibit as uniform a distribution of values across the
range of interest as possible.

It is also necessary if samples from more than one
plant or factory are going to be tested to include a
representative selection of samples from each.

A desirable number of samples for a calibration is
about 40. A regression analysis between the analytical
data and the log values noted for each sample is then
carried out using a Technicon Programme on a Hewlett
Packard calculator. The regression analysis yields the
F values mentioned earlier which can then be entered
into the InfraAlyser to perform future analysis. The
ultimate test of any calibration is on the validity of
results obtained on samples that were not among those
in calibration. It is on this speed, convenience and
validity of the results in use that this paper will
concentrate.

Whole Milk

Whole raw or pasteurised milk is being analysed for Fat
and Total Solids using the liquid cell.

The procedure is to warm the samples to about 45°C in
a water bath. The sample is then placed under the
sampling probe on the external homogeniser unit.
Pressing 'operate' causes the sample to be pumped <u>via</u>
the homogeniser to the temperature-controlled liquid
cell, where measurement takes place.

In the factory situation testing a small number of
samples several times a day may be required.  The labour
involved in carrying out a full bias check or 'daily
calibration check' on each group of samples would not
be justified.  However, in our factory a standard
sample is run with each group of samples and this
detects any instrument drift.  In addition a full
bias check using 10 samples is carried out at least
once a week.

A set of results for comparing the performance of the
InfraAlyser with the Gerber Fat and Density method
for solids is given in Table 1.

The instrument is highly repeatable. Taking two
standard deviations the accuracy is 0.06% for Gerber
Fat and 0.08% for Total Solids.

When testing a number of samples a rate of about 40
samples an hour can be achieved. If a few results are
required quickly so that processing can be started,
bringing the sample to the correct temperature
is the only significant delay.

The technique has the potential of estimating in addition
protein and lactose but this has not to date been
investigated by us.

The InfraAlyser has not been approved for testing <u>ex</u>
farm milk, so a milk buyer could not use it to reject
milk unless the results had been confirmed.

<u>Table 1</u> Comparison of Gerber Fat and Total Solids (Calculated
from Fat and Milk Solids not Fat determined by density)
with Infra Alyser results for Fat and Total Solids

|  | Manual | | Infra Alyser | | Difference Infra Alyser - Manual | |
|---|---|---|---|---|---|---|
|  | Fat | T.S. | Fat | T.S. | Fat | T.S. |
| 1. | 3.70 | 12.53 | 3.68 | 12.27 | -0.02 | -0.26 |
| 2. | 4.10 | 13.00 | 4.10 | 12.79 | 0.00 | -0.21 |
| 3. | 3.70 | 12.43 | 3.64 | 12.22 | -0.06 | -0.21 |
| 4. | 3.75 | 12.55 | 3.70 | 12.30 | -0.05 | -0.25 |
| 5. | 3.85 | 12.67 | 3.85 | 12.42 | 0.00 | -0.25 |
| 6. | 3.90 | 12.77 | 3.86 | 12.40 | -0.04 | -0.27 |
| 7. | 3.95 | 12.82 | 3.95 | 12.64 | 0.00 | -0.18 |
| 8. | 3.70 | 12.46 | 3.69 | 12.19 | 0.01 | -0.27 |
| 9. | 3.80 | 12.64 | 3.78 | 12.34 | 0.02 | -0.30 |
| 10. | 3.60 | 12.39 | 3.59 | 12.15 | <u>0.01</u> | <u>-0.21</u> |
|  |  |  |  |  | <u>-0.01</u> | <u>-0.24</u> |

Standard deviation (n - 1)　　　　　　　　　　0.03　　　　0.04

Control 3.80　12.61　3.74　12.37　　　-0.06　　　-0.24
An example of a single milk repeatability analysed is as
follows:-

|  |  |  |  |  |  |  |  | Mean | Standard deviation |
|---|---|---|---|---|---|---|---|---|---|
| Fat | 3.82 | 3.83 | 3.84 | 3.84 | 3.84 | 3.83 | 3.83 | | 0.01 |
| Total solids | 12.80 | 12.77 | 12.82 | 12.80 | 12.77 | 12.79 | 12.79 | | 0.02 |

Fat 40 - 60% Fat Creams
-------------------------

Control of the fat level in cream is vital for achieving
the correct fat level in double cream and for having cream
of the correct fat level for butter and clotted cream
manufacture.

It was found that creams in the range 40—60% fat will not
always pass through the InfraAlyser homogeniser without
churning.

To overcome this difficulty and get samples successfully
into the liquid cell, it was necessary to disconnect
the feed line from the homogeniser at the point of entry
to the cell.  The disconnected line was placed in a beaker
containing instrument detergent under the sample feed
probe.  When the instrument is operated the detergent
solution is recycled through the homogeniser; while this
is going on it is possible to inject the cream manually
by syringe into the cell.  The InfraAlyser will then
automatically go into the the read operation and when
complete a second syringe is used to wash the cell with
the usual detergent.

Comparisons between the Gerber Fat results and the Infra
Alyser are given in Table 2.

The Gerber test on cream is usually read to only  0.5%
fat.  Thus in some cases the results between duplicates
may vary by by 1%. Against this the InfraAlyser can
be seen to be repeatable and a standard deviation of
0.6% means most results will be within 1% of the Gerber
test.

The repeatability of the InfraAlyser suggests that an
improved calibration may be obtained by calibrating and
testing the calibration against the Rose Gottlieb method.
We have however not been able to do this to date.

Table 2  <u>40 - 60% Fat Cream Infra Alyser Fat results compared</u>
<u>with Gerber</u>

|  | Gerber | Infra Alyser | | Diff. |
|---|---|---|---|---|
| 1. | 41.0 | 39.3 | 39.0 | 1.85 |
| 2. | 41.5 | 40.8 | 40.6 | 0.80 |
| 3. | 45.0 | 44.6 | 44.7 | 0.35 |
| 4. | 48.0 | 45.7 | 47.6 | 0.85 |
| 5. | 48.0 | 47.5 | 47.6 | 0.45 |
| 6. | 50.5 | 49.3 | 49.6 | 1.05 |
| 7. | 55.0 | 55.1 | 55.0 | -0.05 |
| 8. | 59.0 | 58.7 | 58.8 | 0.25 |
| 9. | 61.0 | 60.7 | 60.7 | 0.3 |
| 10. | 61.5 | 61.9 | 61.6 | -0.25 |
|  |  |  | Mean | 0.56 |
|  |  | Std deviation | | 0.6 |
| A) | 37.0 | 37.2 | 36.9 | |
| B) | 37.0 | 37.8 | 37.1 | |
| C) | 56.0 | 56.9 | 56.9 | |
| D) | 57.0 | 57.3 | 56.9 | |

The pairs A + B & C + D were blind duplicates presented
from the same bulk.


<u>Butter</u>

Butter can be analysed for moisture content. An indication
of salt level can also be given, although as salt itself
does not directly affect the infra red this indication
must be achieved by some sort of difference effect.

The InfraAlyser is used by us to monitor freshly churned
butter for moisture content.  The mode of presentation
is to provide a smooth surface in the open cup.

Comparisons of the InfraAlyser with the B.S. Reference
Method and a rapid factory method are given in Table 3.

Table 3  <u>Comparison of Butter Moisture Results by Infra Alyser</u>
<u>Factory Manual and Reference Methods</u>

| Infra Alyser | Factory Manual | Reference | Infra Alyser factory method | Factory Reference | Infra Alyser Reference |
|---|---|---|---|---|---|
| 15.94 | 15.95 | 15.71 | -0.01 | 0.24 | 0.23 |
| 15.78 | 15.81 | 15.54 | -0.03 | 0.27 | 0.24 |
| 15.65 | 15.60 | 15.33 | +0.05 | 0.27 | 0.32 |
| 15.94 | 16.08 | 15.89 | -0.14 | 0.19 | 0.05 |
| 15.79 | 15.79 | 15.54 | 0.00 | 0.25 | 0.25 |
| 15.60 | 15.71 | 15.36 | -0.11 | 0.35 | 0.24 |
| 15.91 | 15.81 | 15.66 | +0.10 | 0.15 | 0.25 |
| 16.16 | 15.92 | 15.90 | +0.14 | 0.02 | 0.26 |
| | | Diff. | 0.00 | 0.22 | 0.23 |
| | | Standard deviation (n - 1) | 0.09 | 0.10 | 0.08 |
| | | Correlation Coefficient | 0.75 | | 0.94 |

The factory manufacturing butter previously used a rapid factory control method performed by heating a dish containing the sample in a Bunsen flame, cooling and reweighing.

During buttermaking, a sample is examined on the Infra Alyser every 20 minutes. Several times a shift, the results from the InfraAlyser are compared with the old factory control method. Examination of Table 3 shows a bias between the reference method and the factory control method but the InfraAlyser has been calibrated

to compare with the factory method. Because of the
narrowness of the range the correlation coefficent
between the two sets of figures has also been calculated.

A correlation is shown between the InfraAlyser and the
factory method. The standard deviation of differences
of 0.09 indicates an expected accuracy within 0.2%
moisture between the two methods on single tests of each
but that an improvement might be achieved between the
butter and the reference method, suggesting some of the
variation is due to the factory method.

Cottage Cheese

A Modern Flexipress enables high volumes of cottage cheese
to be produced. To maintain yield and to comply with
legal requirements moisture control is necessary.

Cottage cheese moistures are being measured in two
locations using the InfraAlyser.

The mode of presentation is to grind the sample to a
smooth paste in a suitable blender. The blender must
operate fairly slowly and be capable of keeping the cottage
cheese in the path of the blades. The ground sample is
presented to the InfraAlyser by forming a smooth surface
in the open cup. The surface is critical in the measurement; we therefore carry out duplicate tests on the same
blend.

A comparison of results is given in Table 4.

The method using the InfraAlyser enables a result to
be obtained within a few minutes of the sample entering
the laboratory. The accuracy compares satisfactorily
with the rapid forced heating method in use previously.
The potential after suitable calibration for doing fat
estimations is also available.

Table 4

Comparison between Cottage Cheese Moistures using the
InfraAlyser and the Reference Method

| Reference results | InfraAlyser | Ref. - InfraAlyser |
|---|---|---|
| 78.5 | 79.4 | -0.9 |
| 78.7 | 79.1 | -0.4 |
| 79.0 | 79.1 | -0.1 |
| 78.8 | 78.9 | -0.1 |
| 79.1 | 78.5 | 0.6 |
| 79.3 | 78.9 | 0.4 |
| 79.2 | 79.3 | -0.1 |
| 79.8 | 79.8 | 0.0 |
| 77.0 | 77.2 | -0.2 |
| | Bias | -0.1 |
| | Standard Deviation of Differences | 0.4 |

## Powder

Skimmed Milk powder is being analysed for moisture content.
Powder is packed into the closed cup for presentation to
the machine.  In our powder factories the InfraAlyser
is used for moisture testing at least once an hour and
comparison is made with reference method 4 hour oven
drying once per shift.  A set of comparisons is given
in Table 5.

Repeatability of both methods is good (Standard deviation
0.04) and the standard differences of 0.08 shows the
accuracy of the InfraAlyser to be usually within 0.2%.

Table 5

Comparisons of Moisture on Skim Milk powder using Infra
Alyser compared to the Oven Drying Method

| InfraAlyser | Reference<br>(Mean of Duplicates) | Difference |
|---|---|---|
| 3.39 | 3.36 | 0.03 |
| 3.66 | 3.65 | 0.01 |
| 3.38 | 3.41 | -0.03 |
| 3.23 | 3.28 | -0.05 |
| 3.39 | 3.38 | 0.01 |
| 3.26 | 3.48 | -0.22 |
| 3.37 | 3.29 | 0.08 |
| 3.23 | 3.20 | 0.03 |
| 3.30 | 3.38 | -0.08 |
| 3.44 | 3.42 | 0.02 |

Correlation   0.78   S.deviation   0.08

Filled Powder

With Filled or Whole Milk powders in addition to moisture
it is possible to obtain fat results at the same time
as moisture.  Table 6 gives a comparison of Gerber fats
on Filled Powder with InfraAlyser fat results.

Gerber fats are read to the nearest 1/3%, and bearing this
in mind the standard deviation 0.6, which will give most
comparisons within 1%, is acceptable.  As with the cream it
may well be possible to improve the calibrations using
the Rose Gottlieb.

Table 6

Comparison of Fat results obtained from the Infra
Alyser with fat obtained by Gerber on Filled Milk Powder

| Gerber | InfraAlyser | Difference |
|--------|-------------|------------|
| 26.4 | 24.9 | 1.5 |
| 29.7 | 29.7 | 0.0 |
| 27.7 | 27.6 | 0.1 |
| 25.7 | 25.8 | -0.1 |
| 26.1 | 25.7 | 0.4 |
| 28.4 | 28.1 | 0.3 |
| 26.4 | 26.3 | 0.1 |
| 30.4 | 29.5 | 0.9 |
| 25.7 | 25.4 | 0.3 |
| 25.4 | 26.4 | -1.0 |

|  | S.deviation | 0.6 |
|--|-------------|-----|
|  | Mean | 0.2 |

Conclusions

Analyses of moisture and fat for a number of dairy products
by near infra red have been discussed.
Once calibration has been achieved rapid results can be
obtained with an accuracy sufficient for process control.

The technique could be extended to protein and carbohydrate
and to products other than those mentioned.

# Measurements of the Principal Constituents of Solid and Liquid Milk Products by Means of Near Infra Red Analysis

By H. R. Egli[1] and U. Meyhack[2]*

[1]CONSERVES ESTAVAYER S.A. (MIGROS), SWITZERLAND
[2]TECHNICON GMBH, 6368 BAD VILBEL I, IM ROSENGARTEN I I, WEST GERMANY

NIR-measurements of milk products become more and more important in the dairy industry. The purpose of this short lecture is to show you what we, as Company Conserves Estavayer, have done in this field.

## 1. Introduction

The InfraAlyzer, properly used, is a very good system. As far as we know it is the only system which is able to analyze a lot of products and constituents without substantial sample preparation. You don't need toxic reagents and the results are available within approximately 2 minutes, e.g. for double measurements. The precision of the system is sufficient for quality control and supervision which is to be seen in the tables later on. Easy handling and built-in diagnostics enables trained persons, e.g. machinists or workers, to run the InfraAlyzer. So laboratory shift-work is not necessary. Furthermore the machinists and the workers have an immediate feedback. Up to now very often standardization, in-process-control and production-control hasn't been possible in the dairy industry. Existing quick-tests very often depend on the laboratory staff, the results are varying and are uncertain too. These gaps are closed by the InfraAlyzer, especially if further developments proposed by ourselves and other users have been fulfilled.

## 2. Survey

### 2.1 Products already calibrated

We are analyzing a lot of products in routine with the
InfraAlyzer (Tables 1,2,3). It is worth mentioning that
it is possible to combine a single product calibration with
a group calibration. We have learned that on choosing the
correct combination of NIR - filters no influence of
colour, cores etc. has been detectable.

Table 1    RESULTS OF ROUTINE MEASUREMENTS:  % DRY MATTER

| Product | Number of Fruits Aromas etc. | N | Measurements R | S |
|---|---|---|---|---|
| Fruit Curd | 4 | 139 | 0,963 | 0,175 |
| Diet Fruit Curd | 3 | 273 | 0,911 | 0,409 |
| Yogurt | 9 | 156 | 0,947 | 0,433 |
| Petit Suisse aux Fruits | 3 | 286 | 0,944 | 0,234 |
| Curd | 1 | 300 | 0,958 | 0,223 |
| Cream Curd | 1 | 263 | 0,970 | 0,263 |
| Caillé de Fromage Frais, 45% Fat in DM | 1 | 300 | 0,973 | 0,135 |
| Caillé de Fromage Frais, 0% Fat in DM | 1 | 300 | 0,903 | 0,183 |
| Skimmed Milk Powder | Rollerdried/ Spraydried | 63 | 1,0 | 0,047 |
| Milk Powder 25% Fat | Rollerdried/ Spraydried | 140 | 0,955 | 0,077 |
| Milk Powder with Vegetable Fat | Spraydried | 227 | 0,966 | 0,078 |
| Condensed Milk | 1 | 291 | 0,893 | 0,266 |

N= Number of Samples
R= Correlation Coefficient
S= Standard Deviation

Table 2      RESULTS OF ROUTINE MEASUREMENTS:   % FAT

| Product | Number of Fruits, Aromas etc. | N | Measurements R | S |
|---------|-------------------------------|---|----------------|---|
| Fruit Curd | 4 | 139 | 0,962 | 0,064 |
| Diet Fruit Curd | 3 | 186 | 0,939 | 0,084 |
| Yogurt | 9 | 138 | 0,996 | 0,072 |
| Petit Suisse aux Fruits | 3 | 283 | 0,905 | 0,283 |
| Cream Curd | 1 | 263 | 0,979 | 0,241 |
| Caillé de Fromage Frais | 1 | 290 | 0,973 | 0,135 |
| Condensed Milk | 2 | 291 | 0,974 | 0,122 |
| Milk Powder 25% | Roller/Spraydried | 140 | 0,936 | 0,089 |
| Milk Powder with Vegetable Fat | 1 | 252 | 0,920 | 0,333 |

N = Number of Samples
R = Correlation Coefficient
S = Standard Deviation

## 2.2 Constituents already calibrated

The most important constituents concerning quality control
of dairy products can be analyzed by the InfraAlyzer
(Table 4). The rapid analysis enables the checking of legal
regulations or internal quality requirements. Corrections
can be done if a product doesn't fit the requirements.
These economic advantages need not to be mentioned.
Finally in this section I want to attract your attention to:
pH-, acidity-, refraction- and density-measurements with the
InfraAlyzer. Analyzing these parameters with NIR is unusual.
We are using the InfraAlyzer for screening-tests, and, if the
result is outside of predetermined bandwidths, classical methods
are used for rechecking. But we can assure you that up to now no
discrepancy has been observed.

Table 3    RESULTS OF ROUTINE MEASUREMENTS: %PROTEIN

| Product | Number of fruits | Measurements | | |
|---|---|---|---|---|
| | | N | R | S |
| Fruit Curd | 4 | 139 | 0,915 | 0,0976 |
| Yogurt | 9 | 108 | 0,984 | 0,0975 |
| Skimmed Milk Powder | Rollerdried/Spraydried | 63 | 0,865 | 0,2047 |
| Milk Powder 25% Fat | Rollerdried/Spraydried | 140 | 0,910 | 0,247 |

RESULTS OF ROUTINE MEASUREMENTS: % LACTOSE

| Product | Number of Fruits, Aromas etc. | Measurements | | |
|---|---|---|---|---|
| | | N | R | S |
| Milk Powder 25% Fat | Rollerdried/Spraydried | 140 | 0,903 | 0,69 |
| Skimmed Milk Powder | Rollerdried/Spraydried | 63 | 0,918 | 0,355 |

N = Number of Samples

R = Correlation Coefficient

S = Standard Deviation

Table 4   CONSTITUENTS ALREADY CALIBRATED   (ROUTINE MEASUREMENTS)

| Product | Dry Matter | Fat | Protein | Lactose | pH | Acidity | Refractivity | Density |
|---|---|---|---|---|---|---|---|---|
| Fruit Curd | X | X | X | | | | | |
| Diet Fruit Curd | X | X | | | | | X | |
| Yogurt | X | X | | | | X | | |
| Petit Suisse aux Fruits | X | X | | | X | | X | |
| Curd | X | | | | X | | | |
| Cream Curd | X | X | | | X | | | |
| Caillé de Fromage Frais, 45% Fat | X | X | | | X | | | |
| Caillé de Fromage Frais, 0% Fat | X | | | | X | | | |
| Skimmed Milk Powder | X | | X | X | | | | |
| Milk Powder 25% Fat | X | X | X | X | | | | |
| Milk Powder with Vegetable Fat | X | X | | | | | | |
| Condensed Milk | X | X | | | | | X | |
| Cream 10-35% Fat | | X | | | | | | X |
| Milk 0-6% Fat | X | X | X | X | | | | |

3. Examples of InfraAlyzer measurements

3.1 Measurements with the Closed cup

3.2 Measurements with the Open Cup

3.3 Measurements with the InfraAlyzer 400 DR

   ( with homogenizer)


Good conformity has been achieved between InfraAlyzer
analysis and manual results concerning all analyzed
constituents (Tables 5,6,7,8,9,10). Only IDF-methods or
methods listed in the " Swiss Food Book" have been used
to calibrate the InfraAlyzer.
In my opinion, there are two reasons for the partly
" bad correlation coefficients" at the testing of the
calibration:

   a)   narrow ranges (there are only small
        variations in our production)
        and/ or
   b)   sometimes the manual results have been
        a little bit smaller or greater than the
        InfraAlyzer results. These are normal
        variations of both methods.

Table 5  MEASUREMENTS WITH THE CLOSED CUP

MILKPOWDER 25% FAT, DATA OF THE CALIBRATION

| | Dry Matter | | Fat | | Protein | | Lactose | |
|---|---|---|---|---|---|---|---|---|
| | Manual | Infra | Manual | Infra | Manual | Infra | Manual | Infra |
| Number of Samples (N) | 69 | 69 | 69 | 69 | 69 | 69 | 69 | 69 |
| Number of Measurements (M) | 138 | 138 | 138 | 138 | 138 | 138 | 138 | 138 |
| Correlation Coefficient (r) | 0,956 | | 0,936 | | 0,910 | | 0,903 | |
| Mean Value ($\bar{X}$) | 96,8 | | 24,7 | | 26,7 | | 36,0 | |
| Range % (R) | 96,3-97,5 | | 24,0-25,3 | | 25,8-28,1 | | 33,7-39,4 | |
| Error of Multiple Determination | * | 0,022 | * | 0,04 | * | 0,123 | * | 0,257 |

*Figures of the manual analysis not available any more

Table 6  MEASUREMENTS WITH THE CLOSED CUP

MILK POWDER 25% FAT, DATA OF CALIBRATION TESTING

| | Dry Matter | | Fat | | Protein | | Lactose | |
|---|---|---|---|---|---|---|---|---|
| | Manual | Infra | Manual | Infra | Manual | Infra | Manual | Infra |
| Number of Samples (N) | 16 | 16 | 16 | 16 | 16 | 16 | 16 | 16 |
| Number of Measurements (M) | 32 | 48 | 32 | 48 | 32 | 48 | 32 | 48 |
| Correlation Coefficient (r) | 0,956 | | 0,857 | | 0,722 | | 0,766 | |
| Mean Value ($\bar{X}$) | 96,78 | | 25,41 | | 25,26 | | 37,18 | |
| Range % (R) | 96,4-97,1 | | 25,2-26,2 | | 24,8-25,6 | | 35,8-38,8 | |
| Error of Multiple Determin. | 0,022 | 0,024 | 0,056 | 0,067 | 0,133 | 0,181 | 0,425 | 0,333 |

Table 7  MEASUREMENTS WITH THE OPEN CUP
FRUIT CURD, 5 TYPES, DATA OF THE CALIBRATION

|  | Dry Matter | | F A T | | Protein | |
|---|---|---|---|---|---|---|
|  | Manual | Infra | Manual | Infra | Manual | Infra |
| Number of Samples (N) | 69 | 69 | 69 | 69 | 69 | 69 |
| Number of Measurements (M) | 139 | 139 | 139 | 139 | 139 | 139 |
| Correlation Coefficient (r) | 0,963 | | 0,962 | | 0,915 | |
| Mean Value (X̄) | 28,5 | | 3,29 | | 7,99 | |
| Range % (R) | 26,5-31 | | 2,9-4,1 | | 7,5-8,7 | |
| Error of Multiple Determin. | * | 0,122 | * | 0,027 | * | 0,060 |

* Figures of the manual analysis not available any more

Table 8  MEASUREMENTS WITH THE OPEN CUP
FRUIT CURD, 5 TYPES, DATA OF CALIBRATION TESTING

|  | Dry Matter | | Fat | | Protein | |
|---|---|---|---|---|---|---|
|  | Manual | Infra | Manual | Infra | Manual | Infra |
| Number of Samples (N) | 69 | 69 | 69 | 69 | 69 | 69 |
| Number of Measurements (M) | 138 | 138 | 138 | 138 | 138 | 138 |
| Correlation Coefficient(r) | 0,964 | | 0,952 | | 0,853 | |
| Mean Value (X̄) | 28,4 | | 3,33 | | 7,95 | |
| Range % (R) | 26,2-31,0 | | 2,8-4,2 | | 7,4-8,8 | |
| Error of Multiple Determin. | 0,106 | 0,105 | 0,029 | 0,034 | 0,052 | 0,066 |

Table 9  MEASUREMENTS WITH THE LIQUID DRAWER AND HOMOGENIZER
CREAM, DIFFERENT FAT CONTENT, DATA OF THE CALIBRATION

| | 12 + 15% Fat | | 18% Fat | | 25% Fat | | 35% Fat | |
|---|---|---|---|---|---|---|---|---|
| | Manual | Infra | Manual | Infra | Manual | Infra | Manual | Infra |
| Number of Samples (N) | 58 | 58 | 29 | 29 | 31 | 31 | 44 | 44 |
| Number of Measurements (M) | 116 | 116 | 58 | 58 | 62 | 62 | 88 | 88 |
| Correlation Coefficient(r) | 0,9992 | | 0,9985 | | 0,9979 | | 0,9971 | |
| Mean Value (X̄) | 14,28 | | 19,2 | | 25,1 | | 33,12 | |
| Range (R) | 9,5-17 | | 15,5-21,5 | | 22-28,5 | | 25-36 | |
| Error of Multiple Determin. | 0,021 | 0,030 | 0,030 | 0,045 | 0,035 | 0,045 | 0,067 | 0,105 |

Table 10  MEASUREMENTS WITH THE LIQUID DRAWER AND HOMOGENIZER
CREAM, DIFFERENT FAT CONTENT, DATA OF CALIBRATION TESTING

| | 12 + 15% Fat | | 18% Fat | | 25% Fat | | 35% Fat | |
|---|---|---|---|---|---|---|---|---|
| | Manual | Infra | Manual | Infra | Manual | Infra | Manual | Infra |
| Number of Samples (N) | 41 | 41 | 25 | 25 | 26 | 26 | 32 | 32 |
| Number of Measurements(M) | 82 | 123 | 60 | 85 | 66 | 97 | 70 | 90 |
| Correlation Coefficient(r) | 0,997 | | 0,967 | | 0,975 | | 0,967 | |
| Mean Value (X̄) | 13,39 | | 18,24 | | 25,24 | | 34,98 | |
| Range % (R) | 11,4-15,8 | | 17,6-20,6 | | 24,1-29,3 | | 34,3-37,0 | |
| Error of Multiple Determin. | 0,025 | 0,026 | 0,030 | 0,049 | 0,073 | 0,088 | 0,037 | 0,052 |

## 4. Examples of ring-tests

The conformity between different methods and laboratories
have been checked, for example several kinds of milk
standardized of different fat content (Tables 11,12).
A similar ring-test has been done for fruit curd
(fat content, Table 13). We assume that the results of
the Roese-Gottlieb-Method are too low, because stabili-
zer and thickening-agents are used during the production
process.

## 5. Some important aspects for the calibration and the routine measurements.

InfraAlyzer measurements are as good as the calibration.
That is why great accuracy and expense has to be made
by calibrating the system. Only products are allowed to
be analyzed which are within the calibration; that means
the conditions for calibrations and for the routine
measurements must be the same.
Statistical tests, delivered with the software, enable
the user to choose the proper filter-combination for his
calibration. Correlation coefficient, standard deviation
and the t-test are the most important parameters for
achieving a good calibration. After developing a calibration,
this calibration must be checked with new, unknown samples.

Assuming a good conformity with the classical methods the
calibration can run for routine measurements.

At Conserves Estavayer, we are making double measurements, whereby the deviations (absolute) are not allowed
to be greater than 0,1 respectively 0,2 for dry matter.

Table 11   METHOD COMPARISON OF DIFFERENT KINDS OF MILK

% FAT: MEAN VALUES OF DOUBLE MEASUREMENTS

| Milk | CESA 1) | | | | | InfraAlyzer | Reference Lab 2) Liebefeld Roese/Gottlieb |
|------|---------|---|---|---|---|-------------|--------------|
|      | Buty Lab 1 | Buty Lab 2 | Mojonnier | Milkotester | Milkoscan | | |
| 1 | 0,05 | -    | 0,12 | 0,07 | -    | 0,08 | 0,07 |
| 2 | 0,57 | 0,48 | 0,61 | 0,54 | 0,66 | 0,64 | 0,55 |
| 3 | 2,22 | 2,13 | 2,18 | 2,08 | 2,14 | 2,20 | 2,13 |
| 4 | 2,80 | 2,62 | 2,76 | 2,75 | 2,75 | 2,77 | 2,72 |
| 5 | 3,14 | 3,15 | 3,10 | 3,02 | 3,09 | 3,12 | 3,05 |
| 6 | 3,45 | 3,45 | 3,45 | 3,44 | 3,46 | 3,44 | 3,41 |
| 7 | 3,82 | 3,59 | 3,76 | 3,71 | 3,74 | 3,67 | 3,70 |
| 8 | 5,17 | 4,97 | 5,05 | 5,13 | 5,05 | 4,99 | 5,01 |
| 9 | 6,30 | 6,25 | 6,18 | 6,22 | 6,16 | 6,12 | 6,05 |

1) CESA = Conserves Estavayer S.A.

2) Liebefeld = Federal Dairy Research Station, Switzerland

Table 12   METHOD COMARISON OF DIFFERENT KINDS OF MILK

MEAN VALUES OF DOUBLE MEASUREMENTS

| Milk | % Protein | | | | % Lactose | | Reference Lab |
| | CESA 1) | | | Liebefeld 2) | CESA 1) | | Liebefeld 2) |
| | Kjeldahl | Milkoscan | InfraAlyzer | Kjeldahl | Milkoscan | InfraAlyzer | IDF-Method |
|---|---|---|---|---|---|---|---|
| 1 | 3,26 | 3,23 | 3,22 | 3,31 | 4,99 | 4,95 | 4,97 |
| 2 | 3,16 | 3,26 | 3,16 | 3,23 | 4,96 | 4,96 | 4,95 |
| 3 | 3,19 | 3,25 | 3,20 | 3,25 | 4,93 | 4,91 | 4,86 |
| 4 | 3,17 | 3,17 | 3,21 | 3,24 | 4,92 | 4,87 | 4,88 |
| 5 | 3,10 | 3,18 | 3,17 | 3,17 | 4,91 | 4,88 | 4,88 |
| 6 | 3,04 | 3,15 | 3,10 | 3,14 | 4,90 | 4,86 | 4,85 |
| 7 | 3,03 | 3,10 | 3,03 | 3,13 | 4,86 | 4,87 | 4,85 |
| 8 | 2,97 | 3,05 | 3,01 | 3,10 | 4,81 | 4,82 | 4,82 |
| 9 | 2,92 | 3,05 | | 3,03 | 4,81 | 4,78 | 4,76 |

1) CESA   =   Conserves Estavayer S.A.

2) Liebefeld   =   Federal Dairy Research Station, Switzerland

Table 13   METHOD COMPARISON

FRUIT CURD (FAT CONTENT)

| Kind | InfraAlyzer CESA 1) | Weibull/Stoldt CESA 1) | Weibull/Stoldt Liebefeld 2) | | Roese/Gottlieb Liebefeld 2) | |
|---|---|---|---|---|---|---|
| Apple | 3,11<br>3,11 | X̄ of 7:<br>3,11 | 3,10<br>3,10 | 3,09<br>3,10 | 3,11<br>3,11 | 3,08<br>3,09 |
| Apricot | X̄ of 5:<br>3,16 | 2,98<br>2,98 | 2,99<br>2,99 | 3,00<br>2,97 | 2,93<br>2,93 | |

1) CESA = Conserves Estavayer S.A.

2) Liebefeld = Federal Dairy Research Station, Switzerland

## 6. Proposals for the improvement of the InfraAlyzer

6.1 A sampler for the homogenizer is absolutly necessary,
as well as a segmented heating-bath for preheating
the samples.

6.2 A mixer at the inlet of the homogenizer is essential
for achieving a good mixing of the sample.

6.3 In-process-control and -correction should be possible
with suitable options.

## 7. Outlook

We think to know the correct filter combination to
analyze fat, protein and dry matter in all dairy products.
In the near future we will realize calculations on a
big computer containing nearly all kinds of dairy products with thousands of results.

# Present and Future of Standard Methods for the Examination of Dairy Products

By G. H. Richardson

DEPARTMENT OF NUTRITION AND FOOD SCIENCES, UTAH STATE UNIVERSITY, LOGAN, UTAH 84322, U.S.A.

Fourteen editions of Standard Methods for the Examination of Dairy Products (SMEDP) have been published by the American Public Health Association since 1905. The 15th edition is due out in 1984. Approximately 5,000 laboratories use SMEDP to assure uniformity in conducting dairy product testing. Originally only bacteriological methodology was emphasized. However, recent editions have included chemical and physical methods to expand utility to the dairy laboratory. The new edition reflects a methods classification scheme that allows meshing with the Association of Official Analytical Chemists Official Methods. A1 and A2 will correspond with Official Final and Official First Action methods, respectively. Class B will be assigned methods that have met all requirements but a collaborative study. New methods being introduced will be given a C classification. Those being phased out will be assigned D while those "grandfather" methods will have an O classification. Methods proposed between editions will be approved by an interim Technical Committee and published in the Journal of Food Protection.

Funding has been provided for the development of SMEDP editions by the US Food and Drug Administration. Future editions will receive more limited support and will not be published as often. With increased international interest and methodology development, the role of SMEDP needs evaluation.

# Advantages and Disadvantages of the EEC System for the Selection of Community Analytical Methods. Preserved Milk and Casein and Caseinates

By G. Vos
COMMISSION OF THE EUROPEAN COMMUNITIES, RUE DE LA LOI 200, B-1049
BRUSSELS, BELGIUM

## 1. Introduction

This paper is especially related to milk products this being
the field better known by the author. However the comments
included may be largely applied to other foodstuffs. Another
point is that the microbiological analysis and health
protection has not been considered as it is a difficult
subject which requires a specific discussion. This emphasises
why this report describes mainly the work performed for
preserved milk and casein and caseinates and the two Council
Directives which were adopted, the first one in 1979[1] and the
second in 1983[2].

## 2. Procedure of adoption of EEC commodity directives

Both of the already mentioned directives include composition
standards which need for their control EEC official methods.
These methods are published in the EEC Official Journal as
Commission Directives (see 3). Council Directives for
commodities such as milk products are included after a certain
delay in the national legislations of each EEC country after
their publication. This occurs at the end of a lengthy
procedure. As a first step, the Commission's services have
to prepare a proposition which requires the consultation of
the Member States, of professional organisations such as
unions, consumer's organisations, industry and the commercial
world. It includes the consultation of the Consulting
Committee for Foodstuffs, where all of these organisations
have a representative. The resulting proposition is then

forwarded to the Council which consults for its own
information the European Parliament and the Social and
Economic Committee (a body which has a composition similar
to that of the Consulting Committee for Foodstuffs). Both
bodies may propose the rejection of the proposition or
amendments which are then followed by a discussion between
the Commission and the representatives of the Member States.
Quite often this can take years before the final adoption
as the agreement of all Member States is required.

3. Official EEC methods adoption

The Commission's Directives procedure for analytical methods
is much faster. The preparatory stage is about the same as
for the Council's Directives (this preparation will be
described in more detail in the next paragraph). After
adoption by the Commission the proposition is sent directly
to the Standing Committee for Foodstuffs which is chaired
by a Commission representative and is adopted by the qualified
majority (usually 8 Member States out of 10) excluding the
unanimity condition.

The acting body concerned in the preparatory work is a working
group of experts originating from the national administration
of the Member States. The Commission's services have contacts
with technical groups of International Scientific Organisations
such as IDF and ISO, professional organisations such as
ASFALEC (EEC Preserved Milk Producers) and the Codex
Alimentairius. When a proposition has been adopted by this
working group it is transmitted for comment to the interested
General Directions (Agriculture, Consumers, etc.), together
with the interested industries. Finally after the Commission's
agreement, it is, as already explained, sent to the Standing
Committee for Foodstuffs.

4. Example of EEC official methods

4.1 Preserved milk for human consumption[3]

The methods are:

- dried matter in evaporated milk (dehydration with sand at
                    99°C);

- humidity in dried milk (drying oven at 102°C ± 1°C);

- fat content in evaporated milk (Rose Gottlieb method);

- fat content in dried milk (Rose Gottlieb method);

- sucrose content in evaporated milk (by polarimetry);

- lactic acid and lactates in dried milk (para-hydroxy
                    diphenyl   colorimetry);

- phosphatose activity in dried milk (modified Sanders and
                    Sayer method);

- phosphatose activity in dried milk (Aschaffenberg and
                    Muller method);

The following comments should be made:

a) all of these methods were proposed by the technical
   committee of ASFALEC;

b) the first six are based on ISO/IDF standards;

c) two methods for the phosphatase activity detection were
   selected as no agreement could be reached regarding a
   choice;

d) the repeatability (agreement between results in the same
   laboratory for the same sample) is mentioned but not the
   reproducibility (agreement between different laboratories);

e) other methods are still to be added in the near future
   (sampling, solubility, etc.)

## 4.2  Edible Caseins and Caseinates

As this commodity Council Directive has only recently been
adopted a corresponding Commission's Directive has not yet
been proposed for the necessary analytical methods.

Table 1 gives a summary of the standards appearing in the
annexes of this directive:

Table 1

(bw = by weight)

| Standards / Products | Maximum moisture content | Minimum milk protein | Minimum casein content | Maximum fat content | Maximum titrable acidity | Minimum or Maximum ash content | Maximum anhydrous Lactose | Sediment content /25g | pH | Maximum Pb Content | Scorched particles - 125g |
|---|---|---|---|---|---|---|---|---|---|---|---|
| Acid Casein | 10 % bw | 90 % bw | 95 % | 2.25 % bw | 0.27 of 0.1 N NaOH | 2.5 %[a] bw | 1 % bw | 22.5 mg | - | 1 mg/kg | nil |
| Rennet Casein | 10 % bw | 84 % bw | 95% | 2.0 % bw | - | 7.5 %[b] bw | 1 % bw | 22.5 mg | - | 1 mg/kg | nil |
| Caseinates | 8 % bw | 88 % bw | | 2.0 % bw | - | | 1 % bw | 22.5 mg | 6.0 to 8.0 | 1 mg/kg | nil |

a Maximum ($P_2O_5$ included)

b Minimum ($P_2O_5$ included)

The preparation of the adoption of this directive began 15 years ago in cooperation with IDF working groups. Several meetings between the EEC and the IDF experts were organised at IDF headquarters in Brussels for the selection of analytical methods. Due to the time needed to reach an agreement for the Council Directive, the services of the Commission decided to limit the extent of the work related to caseins and caseinates by giving priority to the preserved milk methods.

Nevertheless, a first directive has already been provisionally adopted by the working group and it includes the following methods:

- moisture content in acid caseins, rennet caseins and caseinates (drying oven at 102° $\pm$ 1°C);

- protein content of acid caseins, rennet caseins and caseinates (Kjeldahl method with copper sulphate as catalyst);

- free acidity in acid and rennet caseins (extraction by water at 60°C and titration against sodium hydroxyde);

- fixed ash content of acid caseins, ammonium caseinates, whether or not mixed with rennet casein and caseinates, caseins of unknown origin (incineration at 825 $\pm$ 25°C in the presence of magnesium acetate);

- ash content of rennet casein and caseinates except ammonium caseinates (incineration at 825 $\pm$ 25°C);

- pH of caseinates (pH of an equeous extract)

Some comments should be added:

a) all the methods are equivalent to the ISO/IDF standards;
b) only repeatability is mentioned for each method;

As can be seen from Table 1 a certain number of standards are not yet included in this proposition:

- casein content: discussion is under way in the IDF and
  EEC working groups for the whey protein determination
  by colorimetry, photography, split off by mercaptoethanol,
  etc;

- ammonia determination: EEC working group is waiting for
  ISO/IDF working group results;

- fat content: a method has already been provisionally
  adopted by the EEC working group (méthode Schmid -
  Bondzinski-Ratzlaff) similar to the ISO/IDF standard;

- lactose content: same situation as for the fat content
  (sulphuric acid/phenol photometric method);

- sediment content: the EEC working group was planning to
  adopt the ISO standard;

- scorched particles: same situation as for sediment content;

- solubility: the EEC working group has considered the
  possibility of developing a method for the alkaline
  caseinates;

- lead: the EEC working group tried different techniques
  and finally decided to wait for the ISO/IDF developments;

- technical sampling: a proposal for sampling preserved milk
  and casein and caseinates was considered by the EEC working
  group. It is based on the IDF standard;

- sampling with statistical aspect: this difficult problem
  is still under discussion.

## 4.3 Feedstuffs

An EEC scheme directive[4] is the legal base for adoption of
EEC control methods. A number of these methods were
already published in the Official Journal. Some of the
related products are milk products; however, for practical
reasons they use their specific methodology.

## 5. Advantages of the EEC procedure for adoption

The advantage of having EEC official methods for the

application of community directives and regulations is evident
as it greatly reduces the risk of dispute. The methods are
agreed by EEC professional organisations (ASFALEC) and come
from international scientific organisations (IDF, CODEX).
When the standards are correctly stated the application of
the methods limit the risk of accepting a low quality product.
The working group has frequently tested the method by a
round-robin test in order to check its repeatability and
reproducibility.

As the participating laboratories do not always have the
same experience, the test gives the opportunity to examine
the robustness of the method. Sometimes a method is
absolutely necessary as the standard is directly related
to the methadology of measurement (scorched particles for
example).

## 6. Disadvantages of the EEC procedure

As has been mentioned this procedure requires a rather long
delay in preparation due to the number of steps and the
necessity to translate and publish it into all of the official
languages of the EEC. The situation is made even worse when
a recognised international standard does not exist. This
occurs especially in fields where an international
organisation such as IDF is not acting. Due to this official
publication any modification can take quite a time.

Another drawback with this procedure comes from the fact
that a method has to be adopted on a legal base such as a
Community Directive.

In the case of disagreement between states on the accepted
level of an additive an agreed method should help to find
a solution; for example, the normally present level of
nitrate in Cheese is difficult to fix as there is no accepted
Community method. The IDF working group is now studying
this problem.

Last but not least, the selection of a mandatory sampling scheme is not an easy task, and as it is well known, its influence on the obtained methods can be very high.

## 7. Discussions and conclusions

Various proposals may be advanced in order to obtain a better and faster solution. One procedure that could improve the situation is the adoption of a general scheme Directive for analytical methods for foodstuffs similar to the one existing for feedstuffs. This procedure could solve problems such as that already mentioned for nitrate in cheese. However the negative aspects related to the length of the procedure and the selection of a method including its statistical parameters (accuracy , repeatability, reproducibility, etc.) should remain.

When a Commodity Council Directive is under preparation, it is always possible to begin at the same time the preparation of the related analytical method. However, this anticipates the final content and the adoption of the proposal by the Council. The situation with Casein was rather a lucky case as the preparation began in 1971 and the Commodity Directive was adopted this year. Other works performed for commodities such as ice cream, dressings, etc., were not quite as successful because the Council Directives were never adopted.

A second way to solve this question could be the reference to existing standards but some of the standard tests need to be adapted to the EEC requirements for publication.

An interesting way of approach could be found in the COST framework. COST is an agreement of technical and scientific cooperation "a la carte" between the Community and the other Western European countries of OECD. Two examples, COST 90 and 91, exist already in the field of Food Technology. These two concerted actions have to ensure cooperation between the

research programmes of the participating states (12 for COST
90 and 14 for COST 91) in the fields of physical, nutritional
and qualitative properties of foodstuffs.

Various advantages could be found in this kind of cooperation
if it was applied to control methods:

a) limited translation expenses as the official COST
   languages are only French, English and German;

b) possibility to incorporate countries of Western Europe
   who have a high technical level but are not members of
   the EEC;

c) only a small budget is required for concerted action
   programmes, the main part being used to cover secretarial
   and travel expenses;

d) methods could be recommended for use by giving a
   provisional adoption in order to try out the method;

e) there is flexibility as it is always possible for the
   management committee to change the programme of the
   working groups.

[1] OJ No. L 24/43 of 30.1.76 (76/118/EEC)

[2] OJ No. L 237/23 of 26.8.83 (83/417/EEC)

[3] OJ No. L 327/29 of 24.12.79 (79/1067/EEC)

[4] OJ No. L 170/2 of 3.8.70 (70/373/EEC)

# The Selection of Methods for Development as International Standard Methods of Analysis

By R. S. Kirk
LABORATORY OF THE GOVERNMENT CHEMIST, CORNWALL HOUSE, STAMFORD STREET,
LONDON SEI 9NQ, U.K.

When during the last century laboratory work began to be used widely in the
support and control of the manufacture and trade in food products,
background experience in food analysis was sparse.  Analysts soon commenced
to form together into groups for mutual support and to seek standardisation
of test methods and procedures.  The chemists in various food trades
formed specialist national trade associations and federations, many of
which or their offspring are still very active today.

At about the same time, widespread publicity about inferior and adulterated
food brought about national or state legislation to control it.  In the
UK the first Act appeared in 1860[1] and introduced the concept of the local
official enforcement officer, the public analyst.  These analysts also
found it expedient to group together in 1874 into an association, the
Society of Public Analysts[2], and to commence standardisation on methodology.
In the USA the Association of Official Analytical (then Agricultural)
Chemists was formed in 1884 from workers in state, provincial and federal
regulatory agencies and official standard methods began to be published
in 1907.

As trade in food materials and manufactured foods between countries
increased, international trade federations were formed to reduce barriers
to trade.  The preparation of international standard methods of analysis
became a part of the work of several of them, eg. IDF, IOCC, FOSFA,
ICUMSA and ICC.

Government supported national standardisation bodies such as BSI, AFNOR,
DIN and NNI also produce standard test methods to assist their industries.
The move amongst these bodies to produce common international standards
came in 1926 when the International Federation of National Standardising
Associations was formed, to be succeeded in 1947 by ISO.  With a similar
impetus but more from professional scientific interests arose another
international standardisation body, IUPAC.  Also, arising from the

formation of FAO and WHO, the Codex Alimentarius Commission[3] composed of
national government officials was established in 1963 to try to facilitate
international trade in food, to protect the health of consumers and to
prevent fraud by producing standards for the quality and composition of
various foodstuffs and for test methods.

All these different types of bodies and organisations maintain an ongoing
involvement in the preparation of standard test methods.  Their various
objectives and allegiances produce methods which may not be similar,
compatiable or equal in accuracy, precision and sensitivity.  Independent
method development and especially collaborative study may lead to wasteful
duplication of effort.  Different priorities may lead to the retention of
obsolete and outdated official methods which do not provide the analytical
information now needed arising from the great advances in food technology.
Policies may vary on whether modern instrumental methods should be made
official and standard, or whether standard methods should only be simple
procedures capable of use in the small and basically equipped laboratory.

Although there is a move towards rationalisation in the production of
international standard methods, as evidenced by the IDF/AOAC/ISO
collaboration, the duplication of non-equivalent standard methods can
still be a problem for the working analyst and for legislation-producing
bodies such as the EEC.  An example of the problem is the case of the
methodology for the important determination of milk fat content.

The determination of the milk fat (butterfat) content of a product
containing other fats is often required in general food analysis,
expecially in the analysis of chocolate confectionery, but is of no
great importance to a dairy analyst because the products he examines are
in the main purely of dairy origin.  He is more interested in suspected
adulteration of his dairy fat with foreign fats.  The IDF have therefore
produced over the years only standard methods for detecting vegetable
fats[4-6] and the classical Reichert-Polenski empirical volatile acids
distillation method for testing the authenticity of butterfat.  The
semi-micro version of this distillation, originating from an IOCC
standard[8] was adopted in 1974 by the EEC in the absence of any better
international method as the official method for the determination of milk
fat in mixed fats for Common Customs Tariff purposes[9].  This method is of
limited or little use because of its empirical nature and its low
sensitivity and precision.  Several papers had been published by various
authors in the last 20 years on the gas chromatographic separation of

fatty acid methyl, ethyl, butyl or propyl esters[10-15] and in 1976 IUPAC[16] published a comprehensive and generally applicable standard covering a procedure for the gas chromatographic determination of fatty acid methyl esters, including milk fatty acid esters. For analysts wishing to determine only butyric acid as the methyl ester as an indicator of milk fat content, the method was too lengthy because all the esters are chromatographed.

In 1968 Phillips and Sanders[17] described a relatively simple procedure in which only the prepared water soluble fatty acids, *ie.* butyric and caproic acids, obtained from milk fat are directly chromatographed. This method has found extensive use in the Laboratory of the Government Chemist in the UK for more than 12 years and a collaborative study in the UK has been reported with acceptable results[18]. This technique gives the concentration of combined butyric acid in the oil or food product from which can be estimated the milk fat content. The outdated Reichert index, however, is empirically defined and is not equivalent to the butyric acid content. The methods therefore cannot be used alternatively. Yet strangely, IOCC, whose members need more than most an effective method, have not considered replacing their ineffective standard semi-micro Reichert method by a gas chromatographic method. In fact prior to 1983 none of the major standard organisations, apart from one working group of the EEC, have received or made proposals for studying the butyric acid GC method. However, in the regretted absence of any interest or direction from ISO, the IUPAC Commission on Oils, Fats and Derivatives has now set up a working group to study methods of analysis for butyric acid. IOCC have been invited to collaborate and IDF have expressed the wish to be involved. The two current official final action AOAC methods for volatile fatty acids are outdated, lengthy, non-specific and insensitive non-gas-chromatographic procedures[19-20], and accordingly AOAC are expected to collaborate in the IUPAC study.

Butyric acid may not necessarily be the best approach for estimating butter content, so perhaps interested parties should be directed by ISO also to consider the development of alternative indices such as cholesterol or triglyceride ratios.

This paper therefore makes a plea for a more organised, uniform and rational approach to the selection, development, testing and publication of official and standard methods of analysis in the food sector.

1.    There should be more communication between trade associations and
      standardisation organisations on the availability and quality of,
      and requirements for standard methods.

2.    Obsolete methods should be replaced as soon as practicable by more
      modern and efficient methods.

3.    As far as possible methods for particular food trade uses should be
      converted into official ISO procedures or their equivalent for wider
      use and applicability.

4.    There should be a move to rationalise official or standard methods
      from different sources, wherever possible.

A possible protocol to overcome the problems outlined is that where there
is a specific need for a new method, updating or replacing or rationalisation
of a method, the interested party or association should approach ISO
whereupon that body instructs or requests an appropriate body or
organisation to prepare a draft new standard method or procedure.  The method
should then be collaboratively studied by interested parties and subsequently
published as an ISO standard approved and recommended by other involved
organisations as, for example, many IDF/ISO standard procedures now are.

REFERENCES

1.    An Act for Preventing the Adulteration of Articles of Food and Drink,
      23 and 24 Victoria, C.84 (1860).

2.    R C Chirnside and J H Hamence (1974), The 'Practising Chemists',
      pp5-12.  The Society for Analytical Chemistry, London.

3.    G O Kermode (1976).  In 'Food Quality and Safety : A Century of
      Progress', pp185-193.  HMSO, London.

4.    International Standard FIL-IDF 32:1965.  Detection of vegetable fat
      in milk fat by the phytosteryl acetate test; International Dairy
      Federation.

5.    International Standard FIL-IDF 38:1966.  Detection of vegetable fat
      in milk fat by thin layer chromatography of steryl acetates;
      International Dairy Federation.

6.  International Standard FIL-IDF 54:1970. Detection of vegetable fat in milk fat by gas-liquid chromatography of sterols; International Dairy Federation.

7.  International Standard FIL-IDF 37:1966. Determination of soluble and insoluble volatile fatty acid values of milk fat; International Dairy Federation.

8.  Analytical Method 81-E/1960. Determination of the semi-micro values; International Office of Cocoa and Chocolate.

9.  Regulation (EEC) No. 924/74; OJ No. L111, 24.4.74, p.1.

10. L M Smith, J Dairy Sc, 1961, 44, 607.

11. F E Luddy, R A Barford, S F Herb, P Magidman, J Amer Oil Chem Soc, 1968, 45, 549.

12. S Anselmi, L Boniforti, R Monaccelli, Bolletino dei Laboratori clinici provincali, 1960, II, 317.

13. M Iyer, T Richardson, C H Amandson and A Boudreau, J Dairy Sc, 1967, 50, 285.

14. M H Grosjeau and A Fouassin, Revue des fermentations et des industries alimentaires, 1968, 23, 57.

15. J L Iverson and A J Sheppard, J Ass Off Anal Chem, 1977, 60, 284.

16. IUPAC Method 2.302. Gas-liquid chromatography of fatty acid methyl esters. In Standard Methods for Analysis of Oils, Fats and Derivatives, 6th edition 1979. Oxford:Pergamon Press.

17. A R Phillips and B J Sanders, J Ass Publ Analysts, 1968, 6, 89.

18. B O Biltcliffe and R Wood, J Ass Publ Analysts, 1982, 20, 69.

19. Soluble and Insoluble Volatile Acids (Reichert-Meissl and Polenske Values) Official Methods of Analysis of the AOAC, 13th edition, 1980, p442.

20. Mole Per Cent Butyric Acid in Fat, Chromatographic Method. Official Methods of Analysis of the AOAC, 13th edition, 1980, p442.

# Vitamins: An Overview of Methodology

By M. J. Deutsch
FOOD AND DRUG ADMINISTRATION, 200 C STREET, S.W., WASHINGTON D.C. 20204, U.S.A.

The assay of nutrients is a reflection of societal concerns pertaining to the well-being of the population. Initially, the nature of nutrient substances was determined by biological response. As knowledge increased, microbiological, chemical, and physical approaches were used for quantitation. Many of these assays are time consuming. The influence of sample matrix, degradation products formed during processing, and level of analyte in the final product has presented analytical problems.

Biological significance, government regulation, and cost of laboratory methodology have set the priority for this effort. Instrumentation and automation are being implemented because of high professional labor costs and ever-increasing demands. Computer processing of data is a necessary corollary to automation.

The advent of automation and computerization may lead to obsolescence of existing methods of analysis, and this era of rapid change affords a rich field of investigation into more convenient and less variable assays.

A frequent evolution of methods development is the application of the intended assay to pure reference standards, then to single component preparations, then to pharmaceuticals with controlled matrices, and finally to foods or feeds. The diversity of substances consumed as foods is enormous. The influence of excipients in the assay of one food may not be manifested in the matrix of another food. Orange juice is different from milk. While universal applicability is a much desired goal, it is nevertheless frequently elusive. It is, therefore, incumbent on the investigator to define the limits of applicability and, if necessary, exclude certain food groups. Evaluation of methods using sound statistical designs is a science in itself. A chapter could be written about misuses of internal standard techniques but suffice it to say that percent recovery of standard or "spike" recovery may cover a multitude of sins in food assay evaluations.

Rather than address the assay of individual vitamins, which will be covered in contributed papers, I will discuss general problems and considerations which make this an exciting and rewarding field of study.

## Sampling

A universal and frequently overlooked problem is the representativity of the sample aliquot drawn for analysis. The need for compositing large samples has been emphasized in the U.S. regulations pertaining to nutrition labeling. This regulation requires that the sample composite be prepared from twelve retail units of the same production lot drawn at random.

Size of sample, however, is not the only consideration. Comminution, blending, mixing, etc. must be performed by appropriate means to ensure not only that the analyzed portion is representative of the whole, but also that losses due to oxidation, heat, and photolysis are minimized. Seemingly homogenous dairy products such as milk-based infant formulas can be a trap to the unwary analyst and cannot always be eliminated by merely heating a liquid or dissolving solids. Segregation can and does occur in these materials and large sampling errors may result.

## Extraction

Extraction of vitamins from food matrices prior to analysis is the last bastion of manual manipulation in sample assay procedures. Little has been done to automate this step. It is an especially formidable problem when the sample array consists of a variety of matrices and levels of analyte. In addition, the organic solvents used for fat-soluble vitamin extraction have been limited by an increasing awareness of their toxicity or carcinogenicity. Benzene, chloroform, carbon tetrachloride, dioxane, dichloroethane, etc. are examples.

It is in the extraction procedure using wet chemical methods that we perhaps most greatly perturb the structural identity of the organic analytes in a deliberate effort to maximize response. This may produce a misleading perspective on biological activity. The use of strong acids and bases to liberate vitamins from their coenzyme or protein-bound forms does not have a physiological counterpart.

Bioavailability has long been of interest in regard to inorganic nutrients such as iron and calcium. It is becoming of greater concern in regard to organic nutrients. Lysine is the only organic nutrient for which official A.O.A.C. availability methodology has been established, and it is prescribed only for feeds. This is a chemical assay based on determination of the free-epsilon amino groups. Bioavailability is generally determined using the intact animal system with the presumption that results can be extrapolated to the human. It is only with the use of Tetrahymena pyriformis for evaluating amino acids that extensive studies have been performed to correlate microbial response to human response.

In vitamin microbial assays, the sample is often subjected to the same extraction procedure as that used in wet chemical assays because the test organisms are frequently unable to use the bound or complex forms of the vitamin. Most of the organisms used for the determination of biotin are unable to use the bound forms. Loss of free biotin has been postulated as occurring when skimmed milk is dialyzed to reduce its salt content or whey is demineralized. The inclusion of these proteins in infant formulas has led to concern that the biotin present, although released by acid hydrolysis, is of limited biological value to the infant.

An analogous situation exists with respect to niacin, which to a large extent exists in cereals in a chemically bound form. Resolution of niacin bioavailability is complicated by the utilization of dietary tryptophan by mammals as a precursor of nicotinamide nucleotides.

## Multiplicity of Vitamin Forms

The biological activities of essentially all vitamins are manifested by more than one chemical entity. Those which have been synthesized in vitro and added to foods for enrichment or fortification generally do not present analytical problems, mainly because fortification levels are so much greater than levels of nutrients naturally present.

An example is thiamine mononitrate, a synthetic form of vitamin $B_1$, which is the preferred compound under certain matrix conditions. The nitrate is hydrolyzed during acid extraction of the vitamin and thus the thiochrome quantitated is identical to that produced from the thiamine chloride hydrochloride standard. On the other hand, some additives may present analytical problems.

Calcium pantothenate may be added to foods as a racemic mixture. Microbiological assays are responsive to the biologically active form; however, chemical assays which do not include differentiation of isomeric forms can result in grossly erroneous data. In the U.S., erythorbic acid, a geometric isomer of l-ascorbic acid, is a permissible food additive which is practically devoid of antiscorbutic activity. Correct quantitation of the vitamin is mandatory. Other instances exist where the synthetic form may give the same response during analysis as the natural form although their biological activity differs.

The situation becomes much more complex in dealing with naturally occurring analogues. Among the specific cases in point are folates and carotenoids, both of which are represented in nature by a myriad of forms. Some of the problems that arise are related to:

- level of analytical sophistication
- lack of reference standards
- definition and assignment of activity

Separation procedures, for example, GC, HPLC, or other methods capable of distinguishing analogues such as differential growth stimulation of microorganisms, are a credit to the analytical chemist. Extrapolating quantitative data of vitamin analogues and precursors to a single result against which a label claim for nutritional value can be compared presupposes that biological activities of all forms are known. This is often not the case. In several instances, the level of analytical sophistication which has developed far outstrips our capability to use these data fully on a nutritional scale.

In using methods such as HPLC in which analogues are separated, it is mandatory that response be quantitated against an established reference standard. The complexity of this problem is exemplified by the fact that there are over three hundred carotenoids that occur in nature and any that contain the retinol structure have a potential for vitamin A activity. While capturing an eluant fraction to serve as the basis of authenticity is feasible, it nevertheless does not have the validity of a standard whose potency is established by interlaboratory study.

One approach has been to define potency in terms of analytical response. This has been followed by the U.S. Nutrition Research Council and was subsequently used by FDA in establishing the U.S. Recommended Daily Allowances (U.S. RDAs). Folacin is defined empirically as those forms giving a growth response with L. casei; response to folate forms may be significantly affected by selection of conjugase treatment, assay pH, etc. Some investigators have questioned the validity of folic acid activity entries in U.K. food composition tabulations. The same reasoning undoubtedly would pertain to U.S. data. A fruitful area of investigation would be in the use of Tetrahymena gelli for assay of folates since this protozoan has endogenous conjugates.

Foods

The Federal Food, Drug, and Cosmetic Act defines a food as any article that is used as food or drink for man or other animals or components of such articles. The variety and diversity of products present a challenge to the vitamin analyst. Some of the specific areas of concern are:

- fabrications, processing, and preservatives
- influences of other constituents
- nutritional interactions

Many foods in the marketplace are fabricated, that is, composited from natural constituents of varying degrees of isolation or processing and with the vitamin components frequently added in a purified form. It is interesting to note that two major infant health concerns which have arisen during my tenure with FDA have involved vitamin $B_6$ in infant formulas. The dominant form of vitamin $B_6$ in milk is pyridoxal. This form, however, is unstable in the presence of substances released from milk protein by heating. In 1952, a liquid infant formula fortified with pyridoxal was identified as the cause of fifty to sixty cases of vitamin $B_6$ deficiency-induced convulsive seizures. This problem was solved by fortifying the product with the more stable pyridoxol. In the more recent incident the vitamin was inadvertently not added to the product, apparently due to operator error. The latter can occur in any fabrication process but affirmative enforcement activities require more intensive analytical input to prove the absence of a constituent than that required to prove its presence. The pyridoxal-pyridoxol situation exemplified that formulation based on natural composition is not necessarily the preferable course.

In an analogous example, vitamin E assay involves the quantitation of four tocopherols and four tocotrienols. The contribution of ester forms is rare in nature but the gelatin-coated beadlet of the acetate is the usual compound used for fortification of foods and feeds. The free alcohols are added only as antioxidants. Vitamin E acetate can be determined using HPLC methodology that is equally applicable to natural forms. Since fortification levels are higher than those which occur naturally, the assay of esters is relatively straightforward. The only added requirement is that the sample be saponified to convert the ester to free tocopherol.

Whether the added vitamin is present at levels to replace that lost in processing to simulate a natural food source, or is added at levels such as vitamin D in milk fortification is a relevant inquiry. The assay of the latter is facilitated by the large amount of analyte present.

Many HPLC assays have been published for the assay of vitamin D in milk and milk products. Every author is positive that his or her method is the correct method. To date, it seems that the resolution of this situation is still wide open. Three A.O.A.C. collaborative studies are planned in an effort to establish official methodology in this area.

For enforcement purposes, it is pertinent to determine whether methods of analysis developed using food from natural sources are applicable to formulated foods in which major deviations in composition from their natural counterpart may be required.

Quality Assurance

It is essential that adequate quality assurance procedures be included in protocols for vitamin analysis. Concurrent assay of reference standards may suffice for some vitamins but not for others. As an example, calcium pantothenate reference standard may be used to quantitate the amount of pantothenic acid present in a food extract but presupposes that the phosphatase and peptidase enzymes used to liberate the acid from the coenzyme are 100 percent active. One approach to ascertain the efficiency of such treatment is the intralaboratory˙ use of a standard reference material such as Brewer's Yeast. The laboratory would then at least be aware if there are changes in enzyme activity.

Determination of the accuracy of vitamin analyses is more difficult. Frequently, the same result obtained using at least two different methods of analysis based on different principles is presumptive of accuracy. Likewise, corroborating results using official methodology leads to the same conclusion.

Several reference materials in the U.S. have been certified by the government for their mineral content--dried skim milk, orchard leaves, bovine liver, spinach leaves, etc., which are food materials or have characteristics similar to foods. These materials are dried, and considerable effort has been taken to ensure their homogeneity. To date, no such materials are certified for vitamin content although consideration has been given to their production. The primary problem in this endeavor would be that of ensuring that the stability of the organic nutrients in the matrix is maintained for a sufficient time to establish that reliable results are obtained.

The significance of vitamin analysis during the 80's is strengthened by government regulation and recognition of the importance of an informed consumer. Changes in processing technology must also be evaluated to ensure that nutrient content is not thereby significantly diminished. The means of responding to analytical needs are available in instrumental developments.

In some respects, our burdens will be lightened but in other respects they will be more onerous. We must make certain that results we obtain are accurate and that their interpretation is properly integrated with the biological significances of the class of substances designated as vitamins.

# The Chromatographic Analysis of Milk Lipids

By W. W. Christie

HANNAH RESEARCH INSTITUTE, AYR, SCOTLAND KA6 5HL, U.K.

## 1. Introduction

Lipids in milk provide a major source of energy and essential structural components for the cell-membranes in the tissues of the newborn in all mammalian species. They also fulfil these functions when supplied as constituents of other foodstuffs, and can confer distinctive properties on dairy foods that affect processing. For such reasons and because of the commercial importance of milk, milk lipids have probably been studied more intensively than those from any other source. The composition, structures and chemistry of milk lipids have been reviewed on a number of occasions.[1-5] Only thirty years ago, lipids were considered to be oily intractable substances that could be separated into simpler components only with difficulty. This picture was changed by the development of chromatographic procedures, initially gas-liquid chromatography (GLC) and thin-layer chromatography (TLC), and more recently high-performance liquid chromatography (HPLC). These, together with advances in spectroscopic methods, especially mass spectrometry, have brought about the explosive growth in knowledge of lipids in general and milk lipids in particular. Lipid analytical methodology has been reviewed.[6-9]

There are essentially four basic principles that govern the types of separation of lipids that can be achieved by chromatography i.e. partition, adsorption, ion-exchange and complexation. GLC is the best known example of partition chromatography, and has been used to good effect for the analysis of the fatty acid constituents of milk lipids, but has also proved useful for other non-polar aliphatic residues and for molecular species of intact lipids. TLC is the main form of adsorption chromatography in common use, for example, for the separation of individual simple and complex lipid classes.

Although ion-exchange chromatography is not used very often with
lipids, there are some applications to the analysis of complex
lipids, where distinctive separations have been achieved.
Complexation chromatography should not perhaps be considered
apart from the other techniques as it is always used in
conjunction with one or other. As an example, silver
nitrate-impregnated adsorbents are used to effect separations of
lipids according to the number of double bonds in the molecules.
HPLC can also be used in various modes, but mainly adsorption and
reversed-phase partition, to achieve separations, and in essence
uses the same principles as other methods. However, it utilises
recent technological advances in instrumentation and in the
preparation of microparticulate adsorbents and liquid phases to
improve the resolution that can be obtained, and to introduce a
degree of automation into the detection systems, the recording of
separations and the handling of analytical data.

Milk lipid analysis has been a proving ground for many of
these techniques. In this review, the application of
chromatographic procedures will be considered in four areas of
the analysis of milk lipids i.e. the separation and
quantification of lipid classes, the components of lipid classes
(e.g. fatty acids), the positional distributions of fatty acids
in glycerolipids and molecular species of glycerolipids.
Information obtained in this way may be of great value for the
nutritional evaluation of milk products, and to determine those
aspects of the structures of milk lipids that relate to
particular physical properties and consumer applications.

A large number of papers dealing with the analysis of milk
lipids are published every year and it is not possible to review
all of these comprehensively here. Rather, that work which in
the author's opinion is definitive has been described together
with other analyses which may not be unique but appear to be
particularly good examples to illustrate specific points.

2.    Lipid Class Separations
The triacylglycerols are by far the major lipid class in
milk, comprising 97-98% of the total, and they are accompanied by
small amounts only of di- and mono-acylglycerols, free
cholesterol and cholesterol esters (in the approximate ratio
10:1), unesterified (free) fatty acids and phospholipids.

Comparatively large amounts of partial glycerides and unesterified fatty acids have been reported on some occasions, but this usually means that faulty handling of the milk has led to some lipolysis. In most circumstances, TLC would be the method of choice for the analysis of these compounds;[8] they are separated rather easily, but the small amounts of most relative to the triacylglycerols tend to lead to problems in quantification. Charring of the lipids with a corrosive spray followed by photodensitometry of the charred spots is the method used for quantification in most applications, but may not be sufficiently sensitive for the minor constituents of milk fats. A more accurate procedure consists in transesterifying each of the separated lipids in the presence of a suitable internal standard, generally a fatty acid not found naturally in the sample, and then quantifying the resulting methyl ester derivatives by GLC.[10]

More distinctive methods are available for specific simple lipid components. For example, it may be important to monitor lipolysis in milk by measuring the free fatty acid content, and a GLC method has been devised for the purpose.[11] Other sensitive enzymic[12,13] and radiochemical[14] methods have been described but have yet to be applied to dairy products. Lipids containing free hydroxyl groups, present in milk, have been determined by a sensitive spectrophotometric assay following conversion to the pyruvic ester-2,6-dinitrophenylhydrazone derivatives;[15] they include diacylglycerols, hydroxyacylglycerols and sterols. In addition, cholesterol can be determined by a variety of chromatographic and enzymic methods,[8] sterols other than cholesterol in milk have been identified and quantified by GLC,[16-19] but sensitive radiochemical methods must be used for steroid hormones.[20]

The complex lipids in milk have been fractionated by methods analogous to those used for the same lipid classes in animal tissues in general. Phospholipids, for example, can be fractionated by TLC procedures,[8] and related methods have been used for glycolipids including gangliosides,[21,22] although HPLC methods that appear simpler and have greater sensitivity are available for the latter.[8]

Great strides are being made in the development of HPLC methods for the analysis of most lipid classes, but only in a few instances do they appear to have sufficient merit to supplant the older techniques at this stage.

### 3.    Fatty Acids and Related Aliphatic Compounds

A greater range of fatty acids has been isolated or identified as components of milk fats than from any other natural source. For example, a compilation of the fatty acids that had been detected in bovine milk by 1974 listed 437 distinct constituents;[2] they include all the odd- and even-numbered normal saturated fatty acids from $C_2$ to $C_{28}$, monomethyl-branched fatty acids from $C_{11}$ to $C_{28}$ (including positional isomers), multi-methyl-branched fatty acids, monoenoic fatty acids from $C_{10}$ to $C_{26}$ (115 configurational and positional isomers in total), a number of di- and poly-enoic fatty acids, and keto-, hydroxy- and cyclohexyl fatty acids. Many of these are unique to ruminants and are products or by-products of biohydrogenation in the rumen, followed often by further metabolism in the tissues of the animal (*e.g.* by oxidation, chain-elongation or desaturation). With non-ruminants, fewer distinct fatty acids would be expected in the milk, but 183 different components have been identified so far in human milk.[23] Most of these fatty acids have only been detected by using combinations of the more advanced chromatographic, chemical and spectroscopic techniques available. Fortunately, a relative few of these need to be determined when making nutritional evaluations of milk and dairy foods.

GLC is the single most valuable method in use for the analysis of the fatty acid constituents of milk. For this purpose, it is first necessary to transesterify the lipids to more volatile derivatives such as the methyl esters. Because of the volatility and partial solubility in water of the short-chain esters derived from milk fat, the best methods are those with no aqueous-extraction or solvent-removal steps and where the reagents are not heated at any stage. An alkaline trans-esterification procedure using sodium methoxide in methanol, described by Christopherson and Glass,[24] best meets these criteria and has been widely adopted. A related method can be used for lipids in milk other than the triacylglycerols.[25] Although many reseach groups appear to get satisfactory results by transmethylation, it has been argued that more reproducible GLC analyses are obtained if the butyl ester derivatives of the fatty acids are prepared.[26]

With conventional packed-columns and polyester liquid-phases in the gas chromatograph under optimum conditions, up to 53 distinct fatty acids were resolved, identified and quantified in milk fat in a single analysis, for example;[27] these include all the odd- and even-numbered normal fatty acids from 4:0 to 24:0, monoenoic fatty acids from 10:1 to 24:1, and a number of branched-chain and polyunsaturated fatty acids. With the greater degree of resolution attainable by capillary-column GLC, 80 distinct components of milk fat were described in one analysis (although silver-nitrate chromatography was also used here);[28] in addition to those fatty acids mentioned above, many positional isomers of the branched-chain and unsaturated fatty acids were resolved.

While the structure and nature of many of the minor fatty acid components might be considered to have only academic interest, revealing aspects of fatty acid metabolism in the mammary gland and other tissues, those of a few are of real significance. It is important to recognise, for example, that part of the component designated "18:2" or linoleic acid on analysis by GLC may consist of isomers other than cis-9,cis-12 and so may lack biological potency as an essential fatty acid. Indeed, it has been reported that only half the octadecadienoic acid in cows' milk may be the essential isomer.[29] Similarly, a proportion of the monounsaturated fatty acids comprises trans-isomers, the nutritional value of which has been the subject of some debate recently.[30] Methods for the analysis of such compounds are, therefore, of some importance.

TLC with silica gel layers impregnated with silver nitrate has been of particular value in the analysis of the minor fatty acid components. The separation is based on the property that silver salts form polar complexes reversibly with double bonds that retard the migration of the compounds on thin-layer adsorbents:[8,31] saturated compounds do not form complexes so the developing solvent carries them ahead of compounds with one double bond, and these are in turn ahead of compounds with two double bonds and so forth. In addition, trans-double bonds do not form complexes as readily as cis-double bonds, so fatty acids can also be separated according to the configuration of their double bonds. Although model mixtures of cis- and trans-isomers have been separated by GLC on capillary and packed columns containing high-polarity liquid-phases, the complexity of many fat samples, including milk fat, is such that the technique can rarely be of practical value.[8]

Cis- and trans-monoene fractions isolated by silver nitrate TLC contain fatty acids differing greatly in chain-length; a component of a given chain-length can also contain many positional isomers. For further subdivision of monoene fractions according to chain-length, preparative-scale GLC, which can be used with about 0.5mg of material (more with special columns), has generally been the method of choice, although reversed-phase HPLC might now be favoured.[8] With the latter technique, there is no need to volatilise the sample and then condense it following separation for collection purposes, and there need be no losses in the detector.[8,32] It is then necessary to oxidise each of the chain-length fractions, obtaining cleavage between the carbon atoms of the double bonds, and analyse the fragments by means of GLC to determine the positions of those double bonds. Ozonolysis or von Rudloff (permanganate-periodate) oxidation are here the methods of choice, as they cause a minimum of side-reactions.[8,32,33] By using silver nitrate TLC, preparative GLC and oxidative-fission in sequence in this way, the component designated an "18:1" fatty acid in milk fat was found to consist of 4 cis- and 11 trans-isomers, for example.[34] Analogous methods have been used for the di- and poly-unsaturated fatty acids in milk,[35,36] and as mentioned earlier, this may be particularly important for the essential fatty acids.[29]

Useful separations of unsaturated isomers have been obtained by silver nitrate TLC followed by capillary-column GLC, but fewer isomers can be quantified than when oxidative-fission is used.[28,37]

GLC and mass spectrometry in combination can be a very powerful tool and has been used for the tentative identification of many of the unsaturated fatty acids in milk fat.[28] While it reveals the numbers of double bonds in unsaturated fatty acids, it does not necessarily give information on their positions unless they are first converted to appropriate derivatives,[8] and no such application to milk fat appears to have been described. The technique has, however, been used to great advantage for the identification and analysis of the minor branched-chain fatty acids, present in bovine milk, that are not amenable to conventional chemical analysis.[38,39]

Aliphatic compounds contribute greatly to the flavour and palatability of milk and dairy products, and their compositions have been reviewed.[1,40,41] Very many different compounds are involved and a high proportion are derived, chemically or enzymatically, from milk lipids e.g. by hydrolysis and oxidation. Those studied most intensively are lactones and methyl ketones, but short-chain aldehydes and fatty acids are also important. GLC coupled with mass spectrometry has proved particularly useful in their analysis.

The sphingolipids of cows' milk contain a complex range of aliphatic long-chain bases, derived biosynthetically from fatty acids, including normal, iso- and anteiso-saturated di- and tri-hydroxy isomers.[42-44] To analyse these compounds, they were first oxidised to remove the base moiety so that the non-polar aliphatic residues could be identified and quantified by GLC-mass spectrometry.

## 4. The Positional Distributions of Fatty Acids in Milk Triacylglycerols

Triacylglycerols are synthesized in the mammary gland by enzymic mechanisms that exert some specificity in the esterification of different fatty acids at each position of the L- (or sn-)-glycerol moiety. Distinctive structures result that affect the physical properties and digestibility of milk fat. The composition of position sn-2 of the triacylglycerols is perhaps of greatest importance for the consumer, as during digestion by simple-stomached animals and by new-born ruminants, 2-monoacyl-sn-glycerols are formed by the action of the enzyme pancreatic lipase. It appears that such compounds containing relatively-high proportions of palmitic acid, as do those from milk fat, have higher digestibility especially in the new-born.[45,46] On the other hand, the composition of each of the positions is important to the biochemist and may be relevant to the physical properties in consumer applications of milk fat.

The composition of position sn-2 of milk triacylglycerols can be determined by means of hydrolysis in vitro with pancreatic lipase, an enzyme which is almost entirely specific for the primary esters of triacylglycerols. 2-Monoacyl-sn-glycerols are formed that can be separated from the other products of the reaction by TLC before the fatty acid constituents are analysed by GLC.[8,47] For some time, it was thought that there was preferential hydrolysis of the short-chain fatty acids, but it is now known that this is not so although there is indeed some more-rapid hydrolysis of those molecular species of triacylglycerols containing short-chain fatty acids.[48] In practice, the procedure appears to give representative monoacylglycerols and, therefore, satisfactory results.

No lipolytic enzyme has yet been discovered that distinguishes between positions 1 and 3 of a triacyl-sn-glycerol, but a number of ingenious, if complicated, stereospecific analysis procedures have been devised and these have been reviewed.[8,49] Most of the available methods require the preparation of diacylglycerols, which are then chemically phosphorylated for reaction with a stereospecific enzyme, or which are phosphorylated by stereospecific enzymes. The author[8] favours an approach in which $\alpha,\beta$-diacylglycerols (an equimolar mixture of the 1,2- and 2,3-sn-isomers) are prepared by partial hydrolysis of the triacylglycerols with a Grignard reagent, and are converted synthetically to phosphatidylcholines; these are in turn reacted with the phospholipase A of snake venom, which only hydrolyses the "natural" 1,2-diacyl-sn-glycerophosphorylcholine. The products are lysophosphatidylcholine, which contains the fatty acids originally present in position sn-1, free fatty acids released from position sn-2 and the unchanged 2,3-diacyl-sn-glycerophosphorylcholine. Each of these products is isolated by means of TLC and is transesterified for analysis of the fatty acid constituents by GLC. This procedure borrows elements from the work of several research groups.[50-52]

In the analysis of milk lipids by the procedure, problems are encountered in the preparation of pure $\alpha,\beta$-diacylglycerols, because the wide range of chain-lengths of the fatty acid constituents causes band-spreading during isolation by means of TLC, and some contamination with 1,3-diacyl-sn-glycerols results. The first moderately-successful approach[53] utilised $\alpha,\beta$-diacylglycerols prepared by means of pancreatic lipase hydrolysis, but they may not have been entirely representative of those in the triacylglycerols because of selective hydrolysis of the lower molecular-weight species. Others preferred to separate ruminant milk fats by TLC or molecular distillation into a fraction containing predominantly the long-chain fatty acids and one containing the short-chain fatty acids, and to subject these independently to stereospecific analysis, the results being combined later.[54-57] It is also possible, if somewhat less accurate, to use a similar approach, but combine the end-products from hydrolysis of the long-chain and short-chain fractions for fatty acid analysis.[58]

The results of such analyses show that there is a clear preference for palmitic acid to be esterified to position sn-2 of the triacylglycerols, not only in the milk of ruminant animals but also of virtually all other mammalian species.[5,59,60] In ruminant milks, a further distinctive feature is that the short-chain fatty acids, especially butyric and hexanoic acids, are concentrated entirely in position sn-3.[53-58]

The positional distributions of the fatty acids in the main glycerophospholipid constituents of milk, i.e. phosphatidylcholine and phosphatidylethanolamine, have also been determined by means of phospholipase A hydrolysis; these lipids are also asymmetric but not as markedly so as those in many other tissues of the animal.[61-62]

5.    Molecular Species of Milk Triacylglycerols

The complete description of a natural triacylglycerol requires that it be separated into single molecular species in which each position of the glycerol moiety is occupied by only one fatty acid. Patton and Jensen[2] have pointed out that the complexity of milk fatty acids is such that $64 \times 10^6$ individual triacylglycerol species could be present, "the identification of which would keep generations of biochemical taxonomists busy and happy". Even if only the 20 or so most abundant fatty acids are considered, there could be 8,000 molecular species including all positional (and enantiomeric) isomers. It is only realistic, therefore, to attempt to isolate relatively simple molecular fractions rather than single species.

Crude separations into high and low molecular weight fractions are possible by molecular distillation, low-temperature crystallisation and adsorption chromatography. While such procedures may have many practical applications, they tend to have comparatively little analytical value and are not considered further here.

From a practical standpoint, the most valuable analytical procedures are those that enable a rapid "fingerprint" of triacylglycerol molecular species to be obtained so that the physical properties, such as melting or softening points, for example, can be related to definable compositional or structural features. High-temperature GLC can be used for this purpose, and HPLC methods are under development that hold promise for the future.

In high-temperature GLC analyses of triacylglycerols, separation is normally achieved into fractions with the same combined molecular-weights of their fatty acid constituents; the combined chain-lengths of the fatty acyl groups is termed the "carbon number" of the triacylglycerol.[8] No separation according to degree of unsaturation is achieved. Commonly, short glass columns with low levels of non-polar silicone liquid phases are used, as the temperatures (up to $350°C$) required to elute the higher molecular-weight constituents approach the limits of thermal stability both of the stationary phases and of the compounds themselves. Milk fat fractions differing by two carbon atoms from carbon number 30 to 54 have been successfully separated by this means.[56,63] Unfortunately, the presence of odd-chain and branched-chain fatty acids in milk leads to some peak-broadening and loss of resolution. Although there is little margin for error in the preparation of the columns and some preliminary work and practice may be necessary to achieve the required separations and to calibrate for quantitative analysis, this technique appears to be the best available for routine analysis of large numbers of samples.

High-temperature GLC with capillary columns (with glass or fused-silica walls) affords much greater resolution and some preliminary applications to milk triacylglycerols have been described. Components containing odd-chain and branched-chain fatty acids are clearly resolved, and some separation according to degree of unsaturation also appears to be achieved, even with non-polar silicone phases.[64-67] So many components are resolved by this technique indeed, that individual peaks on the recorder trace are not readily identifiable unless access is available to a mass spectrometer.[67,68] Some technical problems remain in obtaining accurate quantification,[64-66,69,70] but the technique will undoubtedly be developed further.

Excellent separations of triacylglycerols have been achieved by means of HPLC in the reversed-phase mode, and some applications have been reviewed.[8,71-74] By far the most widely-used stationary phase consists of octadecyl groups covalently-bound onto microparticles of silica gel, 5 to 10μm in diameter, and it is available under a number of trade names. A considerable effort has been expended in the search for a lipid-sensitive detector for monitoring the eluate from HPLC columns. Differential refractometers, which sense minute differences in the refractive index of the eluate brought about by compounds eluting from the columns, have been used but are not suitable for gradient-elution applications. UV detectors, operating at 200-206nm, have also been applied to the problem but can only be used with a limited range of solvents and these must be of very-high purity. Excellent results in lipid applications have been obtained by using a mass-spectrometer interfaced to an HPLC system as detector,[75] but such equipment is probably too costly for routine use in many laboratories. Also promising are a "mass-detector", which utilises light-scattering following solvent-evaporation as the detection principle,[76-78] and infra-red spectroscopy detectors (at 5.75μm for the carbonyl function),[79-81] although only a few applications to lipids have been described as yet. A system in which the column-eluate was coated onto a moving-wire and passed through a flame-ionisation detector gave excellent results in some hands,[71-74] but was not a commercial success.

Molecular species of triacylglycerols are separated by reversed-phase HPLC according to their partition number, where partition number = carbon number - 2 x (number of double bonds) i.e. a double bond tends to reduce the partition number by 2 units; as separation efficiencies have improved with technical developments, this factor can sometimes be greater than 2. The power of the technique and its potential for milk triacylglycerol fractionation have been shown by two preliminary reports of work in which excellent resolution has been achieved and where the column eluate was monitored by mass spectrometry[75] and mass detection.[78] Further experimental details are awaited with interest. While these appear to be micro-scale separations, suitable for analytical purposes, there is a need for a

larger-scale method so that sufficient material can be obtained
in each molecular fraction for analysis by other methods, and for
determination of physical properties, such as by differential
scanning calorimetry, for example. The most practicable approach
uses HPLC technology adapted to a preparative scale (also in a
reversed-phase mode) with ($C_{15}$ -$C_{18}$)-alkoxypropyl-Sephadex LH20 as
the stationary phase.[82]

For some purposes, it may be necessary to obtain more
information on the molecular structure of a milk fat than can be
obtained by analysis by one of the techniques described above.
Combinations of methods, that make use of different separatory
principles, must then be used in sequence together perhaps with
stereospecific analysis procedures. The most common combination
consists in essence of silver nitrate chromatography (TLC) and
high-temperature GLC, and a detailed protocol for the analysis of
milk fat in this way has been described.[56] In silver nitrate
chromatography in this instance, separation is achieved according
to the total number of double bonds in the three acyl moieties
*i.e.* molecular species with three saturated fatty acids migrate
ahead of those with two saturated and one monoenoic fatty acid,
and so forth. Another time-consuming but fruitful approach has
been reversed-phase column chromatography and pancreatic lipase
hydrolysis of the fractions, by which means 168 molecular species
of the shorter-chain triacylglycerols of milk fat were isolated
and quantified.[83] Related methods have been used by many research
groups for the analysis of milk triacylglycerols.[84-93]

6.   Conclusions
The analysis of milk lipids has stretched many modern
chromatographic methods to their limits. In consequence, we know
more of the detailed structures of milk lipid constituents than
of the lipids from any other natural source. However, the
techniques of capillary GLC and HPLC are being developed rapidly,
and hold promise in terms of speed of analysis, resolution and
convenience for the future.

References
1.    W. R. Morrison, Topics in Lipid Chem., 1970, 1, 51.

2.    S. Patton and R. G. Jensen, Prog. Chem. Fats, 1975, 14, 163.

3.    F. E. Kurtz, "Fundamentals of Dairy Chemistry", 2nd edition, eds. B. H. Webb, A. H. Johnson and J. A. Alford, Avi Publishing Co., Westport, 1974, p. 125.

4.    W. W. Christie, "Lipid Metabolism in Ruminant Animals", ed. W. W. Christie, Pergamon Press, Oxford, 1981, p. 95.

5.    W. W. Christie, "Developments in Dairy Chemistry. 2. Lipids", ed. P. F. Fox, Applied Science Publishers Ltd., London, 1983, p. 1.

6.    W. W. Christie, "Fats and Oils: Chemistry and Technology", eds. R. J. Hamilton and A. Bhati, Applied Science Publishers Ltd., London, 1980, p.1.

7.    A. Kuksis (ed.), "Handbook of Lipid Research", Plenum Press, New York, 1978, Vol. 1.

8.    W. W. Christie, "Lipid Analysis", Pergamon Press, Oxford, 1982.

9.    W. W. Christie and R. C. Noble, "Food Constituents and Food Residues: Their Chromatographic Determination", ed. J. F. Lawrence, Marcel Dekker Inc., New York, in the press.

10.   W. W. Christie, R. C. Noble and J. H. Moore, Analyst (London), 1970, 95, 940.

11.   H. C. Deeth, C. H. Fitz-Gerald and A. J. Snow, N.Z.J. Dairy Sci. Technol., 1983, 18, 13.

12.   K. Mizuno, M. Toyosato, S. Yabumoto, I. Tanimizu and H. Hirakawa, Analyt. Biochem., 1980, 108, 6.

13.   S. Shimizu, K. Inoue, Y. Tani and H. Yamada, Analyt. Biochem., 1979, 98, 341.

14.   H. F. De Brabander and R. Verbeke, Analyt. Biochem., 1981, 110, 240.

15.   H. Timmen and P. S. Dimick, J. Dairy Sci., 1972, 55, 919.

16.   D. R. Brewington, E. A. Caress and D. P. Schwartz, J. Lipid Res., 1970, 11, 355.

17.   V. P. Flanagan, A. Ferretti, D. P. Schwartz and J. M. Ruth, J. Lipid Res., 1975, 16, 97.

18.   A. Adachi and T. Kobayashi, J. Nutrit. Sci. Vitaminol., 1979, 25, 67.

19.   P. W. Parodi, Austral. J. Dairy Technol., 1973, 28, 135.

20.   G. S. Pope and J. K. Swinburne, J. Dairy Res., 1980, 47,

427.

21. R. T. C. Huang, Biochim. Biophys. Acta, 1973, 306, 82.

22. T. W. Keenan, Biochim. Biophys. Acta, 1974, 337, 255.

23. R. G. Jensen, R. M. Clark and A. M. Ferris, Lipids, 1980, 15, 345.

24. S. W. Christopherson and R. L. Glass, J. Dairy Sci., 1969, 52, 1289.

25. W. W. Christie, J. Lipid Res., 1982, 23, 1072.

26. J. L. Iverson and A. J. Sheppard, J. Assoc. Off. Analyt. Chem., 1977, 60, 284.

27. F. Melcher and E. Renner, Milchwissenschaft, 1976, 31, 70.

28. A. Strocchi and R. T. Holman, Riv. Ital. Sostanze Grasse, 1971, 48, 617.

29. K. Kiuru, R. Leppanen and M. Antila, Fette Seifen Anstrichm., 1974, 76, 401.

30. J. E. Kinsella, G. Bruckner, J. Mai and J. Shimp, Am. J. Clin. Nutr., 1981, 34, 2307.

31. L. J. Morris, J. Lipid Res., 1966, 7, 717.

32. J. L. Sebedio, T. E. Farquharson and R. G. Ackman, Lipids, 1982, 17, 469.

33. H. J. Dutton, "Analysis of Lipids and Lipoproteins", ed. E. G. Perkins, American Oil Chemists' Society, Champaign, 1975, p. 138.

34. J. D. Hay and W. R. Morrison, Biochim. Biophys. Acta, 1970, 202, 237.

35. R. G. Jensen, J. G. Quinn, D. L. Carpenter and J. Sampugna, J. Dairy Sci., 1967, 50, 119.

36. H. van der Wel and K. de Jong, Fette Seifen Anstrichm., 1969, 67, 279.

37. P. Lund and F. Jensen, Milchwissenschaft, 1983, 38, 193.

38. A. M. Massart-Leen, H. de Pooter, M. Decloedt and N. Schamp, Lipids, 1981, 16, 286.

39. A. K. Lough, Lipids, 1977, 12, 115.

40. D. A. Forss, J. Am. Oil Chem. Soc., 1971, 48, 702.

41. D. A. Forss, Prog. Chem. Fats, 1972, 13, 177.

42. W. R. Morrison and J. D. Hay, Biochim. Biophys. Acta, 1970, 202, 460.

43. W. R. Morrison, FEBS Letters, 1971, 19, 63.

44. W. R. Morrison, Biochim. Biophys. Acta, 1973, 316, 98.

45. R. M. Tomarelli, B. J. Meyer, J. R. Weaber and F. W. Bernhart, J. Nutr., 1968, 95, 583.

46. L. J. Filer, F. H. Mattson and S. J. Fomon, J. Nutr., 1969, 99, 293.

47. J. Brockerhoff and R. G. Jensen, "Lipolytic Enzymes", Academic press, New York, 1974.

48. J. Sampugna, J. G. Quinn, R. E. Pitas, D. L. Carpenter and J. G. Jensen, Lipids, 1967, 2, 397.

49. W. C. Breckenridge, "Handbook of Lipid Reseach", ed. A. Kuksis, Plenum Press, New York, 1978, Vol. 1, p. 197.

50. H. Brockerhoff, J. Lipid Res., 1965, 6, 10.

51. W. W. Christie and J. H. Moore, Biochim. Biophys. Acta, 1969, 176, 445.

52. J. J. Myer and A. Kuksis, Can. J. Biochem., 1979, 57, 117.

53. R. E. Pitas, J. Sampugna and R. G. Jensen, J. Dairy Sci., 1967, 50, 1332.

54. W. C. Breckenridge and A. Kuksis, Lipids, 1968, 3, 291.

55. W. C. Breckenridge and A. Kuksis, Lipids, 1969, 4, 197.

56. A. Kuksis and W. C. Breckenridge, "Dairy Lipids and Lipid Metabolism", ed. M. F. Brink and D. Kritchevsky, Avi Publishing Co., Westport, 1968, p. 28.

57. L. Marai, W. C. Breckenridge and A. Kuksis, Lipids, 1969, 4, 562.

58. W. W. Christie and J. L. Clapperton, J. Soc. Dairy Technol., 1982, 35, 22.

59. P. W. Parodi, Lipids, 1982, 17, 437.

60. M. R. Grigor, Comp. Biochem. Physiol., 1980, 65B, 427.

61. W. R. Morrison, E. L. Jack and L. M. Smith, J. Amer. Oil Chemists' Soc., 1965, 42, 1142.

62. W. R. Morrison and L. M. Smith, Lipids, 1967, 2, 178.

63. A. Kuksis, L. Marai and J. J. Myher, J. Amer. Oil Chemists' Soc., 1973, 50, 193.

64. G. Schomburg, R. Dielmann, H. Husmann and F. Weeke, J. Chromat., 1976, 122, 55.

65. K. Grob, H. P. Neukom and R. Battaglia, J. Amer. Oil Chemists' Soc., 1980, 57, 282.

66. H. Traitler and A. Prevot, J. High Res. Chrom., Chrom. Rev., 1981, 4, 109.

67. S. G. Wakeham and N. M. Frew, Lipids, 1982, 17, 831.

68.   T. Murata and S. Takahashi, Analyt. Chem., 1973, 45, 1816.

69.   K. Grob, J. Chromat., 1979, 178, 387.

70.   K. Grob, J. Chromat., 1981, 205, 289.

71.   E. W. Hammond, J. Chromat., 1981, 203, 397.

72.   C. H. S. Hitchcock and E. W. Hammond, "Developments in Food Analysis Techniques", ed. R. D. King, Applied Science Publishers, Barking, 1980, Vol. 2, p. 185.

73.   E. W. Hammond, "HPLC in Food Analysis", ed. R. Macrae, Academic Press, London, 1982, p. 167.

74.   K. Aitzetmuller, Prog. Lipid Res., 1982, 21, 171.

75.   A. Kuksis, L. Marai and J. J. Myher, J. Chromat., 1983, 273, 43.

76.   J. M. Charlesworth, Analyt. Chem., 1978, 50, 1414.

77.   R. Macrae, L. C. Trugo and J. Dick, Chromatographia, 1982, 15, 476.

78.   A. Stolyhwo, H. Colin and G. Guiochon, J. Chromat., 1983, 265, 1.

79.   N. A. Parris, J. Chromat., 1978, 149, 615.

80.   N. A. Parris, J. Chromat., 1978, 157, 161.

81.   D. S. Atkin, R. J. Hamilton, S. F. Mitchell and P. A. Sewell, Chromatographia, 1982, 15, 97.

82.   B. Lindqvist, I. Sjogren and R. Nordin, J. Lipid Res., 1974, 15, 65.

83.   L. J. Nutter and O. S. Privett, J. Dairy Sci., 1967, 50, 1194.

84.   I. M. Morrison and J. C. Hawke, Lipids, 1977, 12, 994.

85.   P. W. Parodi, Austral. J. Dairy Technol., 1980, 35, 17.

86.   P. W. Parodi, J. Dairy Res., 1982, 49, 73.

87.   A. A. Y. Shehata, J. M. DeMan and J. C. Alexander, Canad. Inst. Food Technol. J., 1971, 4, 61.

88.   A. A. Y. Shehata, J. M. DeMan and J. C. Alexander, Canad. Inst. Food Technol. J., 1972, 5, 13.

89.   M. W. Taylor and J. C. Hawke, N.Z.J. Dairy Sci. Technol., 1975, 10, 40.

90.   I. M. Morrison and J. C. Hawke, Lipids, 1979, 14, 391.

91.   P. W. Parodi, J. Dairy Res., 1979, 46, 633.

92.   P. W. Parodi, J. Dairy Res., 1981, 48, 131.

93.   R. E. Timms, Austral. J. Dairy Technol., 1980, 35, 17.

# Chromatographic Determination of Vitamins in Dairy Products

By R. Macrae[1], I. A. Nicolson[1], D. P. Richardson[2], and K. J. Scott[3]

[1]DEPARTMENT OF FOOD SCIENCE, UNIVERSITY OF READING, LONDON ROAD, READING, BERKSHIRE RGI 5AQ, U.K.
[2]CARNATION LTD., DANESFIELD HOUSE, MEDMENHAM, MARLOW, BUCKINGHAMSHIRE, U.K.
[1]NATIONAL INSTITUTE FOR RESEARCH IN DAIRYING, SHINFIELD, READING, BERKSHIRE RG2 9AT, U.K.

INTRODUCTION

Milk and related dairy products have traditionally been associated with foods of high nutritional quality, both in terms of macronutrients (proteins) and micronutrients (vitamins and minerals). Of these compounds the determination of vitamins has presented the analyst with the most problems. Furthermore, of the nutrients, the vitamins are the least stable and therefore most likely to be destroyed during storage or processing[1]. Their determination at all stages in the food chain then becomes important. Vitamins of both the fat-soluble group (e.g. vitamins A, D and E) and water-soluble group (e.g. B group vitamins, vitamin C and folic acid) are present in dairy products, but here only the determination of the latter will be discussed.

Chemical, physicochemical and microbiological methods have been adopted for the determination of water-soluble vitamins in dairy products[2]. All of these methods have their strengths and weaknesses, and the decision to use a particular method for a given analysis must take factors such as specificity, sensitivity, accuracy and analysis time into consideration.

The most frequently studied vitamins in dairy products include vitamin $B_1$ (thiamin), vitamin $B_2$ (riboflavin) and the vitamin $B_6$ group (pyridoxine, pyridoxal and pyridoxamine) and, to a lesser extent, niacin/niacinamide. Vitamin C is rapidly destroyed in fresh milk[3] but its determination in fortified products, e.g. baby foods, is important. The methods for vitamin $B_{12}$[4] are restricted to microbiological methods as it is only present at very low levels. The determination of vitamin

$B_1$, vitamin $B_2$ and the $B_6$ group vitamins will be discussed here to illustrate the various methodologies available, together with their relative merits.

Vitamin $B_1$ is present in foods, and indeed in dairy products, at very low levels and adequate sensitivity for its determination is achieved by conversion to a fluorescent derivative (thiochrome). This provides the basis for a standard method[5] but the conditions of oxidation are critical, and as thiamin may be present as the pyrophosphate ester and/or bound to protein[3], a preliminary enzymic hydrolysis is required. Several chromatographic methods have been described but in general they lack the sensitivity to determine thiamin at natural levels[6]. In many HPLC methods thiamin is eluted with poor peak shape, even in the presence of ion-pair reagents, and this also contributes to poor detection limits. Both of these problems can be circumvented, to some extent, by carrying out a precolumn oxidation to produce thiochrome, when fluorescence detection can be used. However, the additional precolumn stage increases the complexity of the analysis and introduces further potential sources of error. This method has been applied to dairy products[7]. The inherent sensitivity of microbiological methods is used to advantage in thiamin assays but the problems of extraction and hydrolysis still remain, although they are reduced as sample amounts can be considerably reduced.

Vitamin $B_2$ with its natural fluorescence presents the analyst with a somewhat simpler problem. The standard method involves direct measurement of fluorescence of extracts with a procedure for eliminating fluorescence due to other components by measurement before and after destroying riboflavin by reduction with dithionite[5]. Riboflavin's natural fluorescence is also used to advantage in liquid chromatographic methods[7-10] where the selectivity of fluorescence detection allows relatively simple extraction and chromatographic procedures to produce a specific method. As with thiamin, riboflavin may be present as phosphate esters, although the majority of reports suggest that in cow's milk riboflavin exists in the free form[3].

Vitamin $B_6$ presents a rather different problem in that it is often present in one or more forms and as the corresponding phosphate esters. No simple chemical procedure exists for the quantification of all three forms after acid and enzyme hydrolysis, and methods based on the conversion of the three forms to a single compound are subject to error introduced by variable conversion rates[11]. An ideal method would therefore determine the three forms individually.

HPLC methods have only recently appeared [12,13] but some have reported variable results[14,15] as a result of interfering peaks. The use of fluorescence detection produces adequate sensitivity for most samples and careful selection of the chromatographic conditions leads to separation of the three main vitamin $B_6$ components.

Microbiological methods provide high sensitivity but unless a "total $B_6$" value is required separation by ion-exchange prior to determination is required. Here again, the existence of phosphate esters complicates the assay as the free and phosphorylated forms can produce different growth responses[3].

The limitations of the various analytical methods are known but the extent to which these limitations are reflected in low precision or accuracy for particular food products has not been adequately investigated. There is a clear need for a detailed comparison of methodologies and this is nowhere more important than in dairy products, especially those used for infant feeding. The present paper presents some preliminary work in which HPLC and microbiological methods have been compared for the determination of vitamins $B_2$ and $B_6$ in a range of dairy products. Vitamin $B_1$ was not included in this comparison because of the insensitivity of the HPLC method when applied to natural levels.

MATERIALS AND METHODS

Equipment

The riboflavin analyses were performed using an ACS Model 750/03 pump (ACS Ltd, Luton, U.K.) with fluorescence detection using a Perkin Elmer Model 3000 fluorescence spectrophotometer (Perkin Elmer Ltd, Beaconsfield, U.K.). The column was a 15 cm x 4.6 mm i.d. Technicon Fast LC-8 (Technicon Ltd, Basingstoke, U.K.).

The pyridoxamine and pyridoxal analyses were performed using a Gilson gradient system with a 704 HPLC system manager

(Anachem Ltd, Luton, U.K.) with fluorescence detection as previously described.    The column employed was a 25 cm x 4.6 mm i.d. Spherisorb ODS-2 (Phase Separations Ltd, Clwyd, U.K.).

The pyridoxine analyses were performed using the same system as in the riboflavin analyses but with the 25 cm ODS-2 column.

All assays used a Rheodyne Model 7125 injection valve (Rheodyne Inc., California, U.S.A.) and a Bryans Model BS600 recorder (Bryans Ltd, Surrey, U.K.).

Reagents

Riboflavin and pyridoxine hydrochloride were donated by Roche Products Ltd (Welwyn Garden City, U.K.).    Pyridoxal hydrochloride and pyridoxamine dihydrochloride were purchased from Sigma Chemical Co. Ltd (Poole, U.K.).    All other chemicals were of Analar or HPLC grade and were purchased from the usual chemical suppliers.    HPLC solvents were purchased from Rathburn Chemicals (Walkerburn, U.K.).

Samples

Samples of the commercial skimmed milk powder were taken from a single production run and subsequently stored under various conditions at the National Institute for Research in Dairying (N.I.R.D.), Shinfield, Reading, U.K.    A selection of infant formulas purchased locally were provided by N.I.R.D.

Determination of vitamin $B_2$ in skimmed milk powder and infant formulas

HPLC analyses.    One ml sample was pipetted into a glass universal bottle with 20 ml 0.1 M HCl.    The bottles were sealed, the contents mixed and then autoclaved for 15 min at 121°C.    After allowing the contents to cool, they were trans- ferred to a 50 ml volumetric flask, made up to volume with distilled water and filtered (Whatman No. 41 and Millipore 0.45 $\mu$m).    The following chromatographic conditions were used:

Mobile Phase:    0.001 M tripotassium citrate, pH adjusted to 3.5 with dil. HCl:methanol (70:30, V/V).

Flow rate:    1.5 ml min$^{-1}$

Detection:    Fluorescence; excitation 450 nm (slit width 15 nm), emission 520 nm (slit width 20 nm), x 20 expansion.

Quantification: Based on peak height.

Microbiological analysis.   All the microbiological assays were undertaken by N.I.R.D. using established procedures.   The samples were hydrolysed using the same conditions as the HPLC analysis[16], and then riboflavin was assayed using Lactobacillus casei (NCDO 243), using a medium based on the method of Roberts and Snell[17].

Determination of vitamin $B_6$ in skimmed milk powder

HPLC analysis.   The samples (5 g) were transferred to 50 ml volumetric flasks with 4 ml trichloroacetic acid (40%, m/V). The flasks were held at $4^\circ C$ for 10 min, made up to volume with distilled water and filtered (Whatman No. 41 and Millipore 0.45 μm).   The following chromatographic conditions were used for the determination of pyridoxamine and pyridoxal:

Mobile Phase:   A.   0.005 tripotassium citrate, pH adjusted to 2.5 with dilute sulphuric acid.

B.   0.01 M tripotassium citrate, pH adjusted to 2.5 with dilute sulphuric acid:methanol (50:50, V/V).

| Gradient: | Time (min) | %B | Time (min) | %B |
|---|---|---|---|---|
| | 0 | 0 | 15 | 50 |
| | 4 | 0 | 17 | 0 |
| | 10 | 10 | 20 | 0 |

Flow Rate:     1 ml $min^{-1}$

Detection:     Fluorescence;   excitation 290 nm (slit width 15 nm), emission 390 nm (slit width 20 nm), x 20 expansion.

Quantification: Based on peak height.

Determination of vitamin $B_6$ in infant formulas

HPLC Analysis.   Pyridoxamine and pyridoxal were determined as in the skimmed milk powder.   Pyridoxine levels were determined in the same extracts using the following chromatographic conditions:

Mobile Phase:   0.01 M tripotassium citrate, pH adjusted to 2.5 with dilute sulphuric acid:methanol (95:5, V/V).

Flow Rate:     1 ml $min^{-1}$.

Detection:     As for previous vitamin $B_6$ determination.

Quantification: As for previous vitamin $B_6$ determination.

Microbiological analysis.    Microbiological assays, carried out
at N.I.R.D., were used to determine total vitamin $B_6$ in all the
samples, using well established techniques.

      The samples (1 ml, 100 mg $ml^{-1}$ were autoclaved in 40 ml
0.055 M HCl for 2 hours at $121^{\circ}C$[18].   After cooling the pH was
adjusted to 4.6-4.8 and the extract made up to volume and
filtered .   Subsequently, vitamin $B_6$ was determined using
**Klockera** apiculata (NCYC 245) using a medium as described by
Barton-Wright[19].

RESULTS   AND   DISCUSSION

Riboflavin

      The same extraction procedure was adopted for both the
HPLC and microbiological assays as the techniques had similar
sensitivities.    This allows a direct comparison of the
analytical data.    A selection of skimmed milk powders, stored
under various conditions, and a range of commercial infant
formulas purchased locally, were studied.    In all cases a very
simple chromatogram was obtained, as a result of the selective
fluorescence detection.    A typical chromatogram is shown in
Fig. 1.

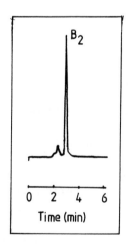

Fig. 1     Determination of riboflavin in dried skimmed milk.

The data from the two assay procedures are collated in Table 1.   The results from the methods appear similar and indeed this is shown to be so by a statistical comparison (significantly different only at 5% level).   In general the HPLC method appears to produce less variation within each sample, but as the replication is different for each method a valid statistical comparison of precision is not possible.

Vitamin $B_6$

The same samples were then examined for vitamin $B_6$.   In the case of HPLC, pyridoxine, pyridoxamine and pyridoxal were determined individually and summed to produce a total vitamin $B_6$ value.   Whereas the microbiological assay simply determined total vitamin $B_6$ directly.   In this case the same extraction procedure could not be adopted, as the HPLC method required considerably more concentrated extracts.   Additionally, hydrolysis was found to produce interfering peaks and therefore was omitted from the HPLC method.   A direct comparison of the data from the two techniques must therefore be treated with caution, as differences may have been introduced at the extraction stage.

HPLC analysis of the three vitamin $B_6$ forms required two sets of chromatographic conditions (see experimental). However, in the case of the skimmed milk powders no pyridoxine was detected and thus only one chromatographic run was required (Fig. 2).

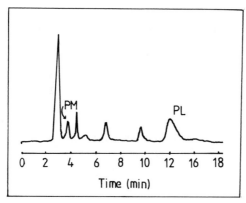

Fig. 2    Determination of pyridoxamine and pyridoxal in dried skimmed milk.

Table 1.   Comparison of HPLC and microbiological assay for the
determination of vitamin $B_2$ (all expressed as mgg$^{-1}$)

| SAMPLE | | | TECHNIQUE | | | |
|---|---|---|---|---|---|---|
| (a) Dried skimmed | HPLC* | | | MICROBIOLOGICAL | | |
| Milk powder | (1) | (2) | (1) | (2) | (3) | (4) |
| 1 | 17.0 | 16.9 | 18.8 | 20.2 | 19.2 | 19.6 |
| 2 | 16.8 | 17.1 | 20.1 | 22.3 | 17.8 | 18.9 |
| 3 | 16.2 | 16.8 | 19.2 | 22.7 | 18.0 | 19.2 |
| 4 | 16.8 | 16.6 | 19.2 | 21.1 | 16.6 | 19.1 |
| 5 | 17.1 | 17.2 | 20.7 | 21.1 | 18.6 | 19.3 |
| 6 | 16.2 | 15.6 | 19.6 | 19.8 | 18.8 | 17.0 |
| 7 | 16.1 | 16.2 | 18.0 | 22.1 | - | 17.5 |
| 8 | 14.5 | 14.9 | 18.0 | 18.4 | - | 15.8 |
| (b) Infant formulas | (1) | (2) | | (1) | (2) | |
| 1 | 1.13 | 1.12 | | 1.18 | 1.32 | |
| 2 | 1.52 | 1.52 | | 1.88 | 1.94 | |
| 3 | 2.58 | 2.47 | | 2.40 | 2.51 | |
| 4 | 1.91 | 1.98 | | 1.93 | 2.28 | |
| 5 | 0.93 | 0.95 | | 0.92 | 1.02 | |
| 6 | 1.12 | 1.10 | | 1.08 | 1.12 | |
| 7 | 1.56 | 1.54 | | 1.54 | 1.70 | |
| 8 | 1.49 | 1.49 | | 1.66 | 1.64 | |
| 9 | 1.59 | 1.59 | | 1.60 | 1.78 | |
| 10 | 1.55 | 1.55 | | 1.48 | 1.52 | |
| 11 | 1.15 | 1.17 | | 1.10 | 1.26 | |

\*   Mean of two analyses

Infant formulas are fortified with pyridoxine and therefore a second chromatographic analysis of the same extract was required (Fig. 3).

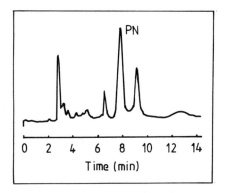

Fig. 3    Determination of pyridoxine in an infant formula.

With the skimmed milk powders (Table 2) it was shown that there was no significant difference between the two sets of data.    Unfortunately with the infant fomulas only one set of microbiological data was available and so a statistical comparison of data was not possible,  but there do not appear to be any major discrepancies.

CONCLUSIONS

Previous comparisons of this kind have involved the analysis of a wide range of different foods, with varying results.    In general, HPLC methods for the determination of riboflavin compare well with standard methods[7,9,10]; this has been confirmed by this set of data.    However, when HPLC methods for the determination of vitamin $B_6$ have been compared, some have produced large discrepancies[16,17].    The results of this comparison of vitamin $B_6$ methodologies have not shown great variability,  despite differences in the extraction procedure employed.    Further comparisons of this kind, with perhaps more emphasis on extraction procedures, are required before it can be concluded that HPLC is a viable alternative to existing methods.

Table  2.  Comparison of HPLC and microbiological assay for
the determintion of vitamin $B_6$ (all expressed as $\mu g g^{-1}$).

SAMPLE                                    TECHNIQUE

(a) Dried skimmed milk powder

| | HPLC* | | | | | | MICROBIOLOGICAL | | | |
| | PM | | PL | | TOTAL | | TOTAL VITAMIN $B_6$ | | | |
| | (1) | (2) | (1) | (2) | (1) | (2) | (1) | (2) | (3) | (4) |
|---|---|---|---|---|---|---|---|---|---|---|
| 1 | 2.3 | 2.2 | 2.7 | 2.6 | 5.0 | 4.8 | 5.1 | 4.6 | – | 5.0 |
| 2 | 1.9 | 2.0 | 2.6 | 2.8 | 4.5 | 4.8 | 5.4 | 4.8 | 4.6 | 4.6 |
| 3 | 2.1 | 2.0 | 3.0 | 2.7 | 5.1 | 4.7 | 5.0 | 5.2 | 4.9 | 4.6 |
| 4 | 2.5 | 2.6 | 3.2 | 3.1 | 5.7 | 5.7 | 4.7 | 5.3 | 4.9 | 4.5 |
| 5 | 2.0 | 1.9 | 2.9 | 2.7 | 4.9 | 4.6 | 5.1 | 5.6 | 4.6 | 4.7 |
| 6 | 1.7 | 1.7 | 2.5 | 2.6 | 4.2 | 4.3 | 5.4 | 5.5 | 4.5 | 5.6 |
| 7 | 1.9 | 1.9 | 2.7 | 2.7 | 4.6 | 4.6 | 5.7 | 5.6 | 4.9 | 5.0 |

(b) Infant formulas

| | HPLC* | | | | | | | MICROBIOLOGICAL TOTAL° | |
| | PM | | PL | | PN | | TOTAL | | VITAMIN $B_6$ |
| | (1) | (2) | (1) | (2) | (1) | (2) | (1) | (2) | (1) |
|---|---|---|---|---|---|---|---|---|---|
| 1 | ND+ | ND | 1.1 | 1.1 | 5.2 | 5.2 | 6.3 | 6.3 | 6.2 |
| 2 | ND | ND | 1.2 | 1.3 | 4.8 | 4.8 | 6.0 | 6.1 | 6.7 |
| 3 | 0.8 | 0.7 | 1.3 | 1.1 | 5.6 | 5.6 | 7.7 | 7.4 | 6.8 |
| 4 | 1.0 | 1.0 | 1.3 | 1.3 | 3.2 | 3.3 | 5.5 | 5.6 | 5.2 |
| 5 | 1.0 | 1.1 | 1.0 | 0.9 | 4.0 | 4.0 | 6.0 | 6.0 | 5.0 |
| 6 | ND | ND | 1.0 | 1.0 | 6.2 | 6.3 | 7.2 | 7.3 | 7.2 |

| * | Mean of two analyses | PM | Pyridoxamine |
| + | ND = None detected | PL | Pyridoxal |
| ° | Mean of four dose levels | PN | Pyridoxine |

ACKNOWLEDGEMENTS

This work was part of a CASE Award with Cadbury Schweppes PLC and the Department of Food Science, University of Reading.

The authors would like to thank Dr D. Collett, Department of Applied Statistics, University of Reading, for his statistical advice.

REFERENCES

1. A.E. Bender, "Food Processing and Nutrition", Academic Press, London, 1978.
2. A.A. Christie and R.A. Wiggins, "Developments in Food Analysis Techniques", R.D. King. (Ed.), Applied Science Publishers, London 1978, Vol. I, p.1.
3. A.M. Hartman and L.P. Dryden, "Fundamentals of Dairy Chemistry", B.H. Webb, A.H. Johnson and J.A. Alford (Eds), Avi Publishing Co. Inc., Westport, 1974, p.325.
4. C.J. Blake, BFMIRA Scientific and Technical Survey No. 95, 1977.
5. W. Horwitz Ed., "Official Methods of Analysis of the Association of Official Analytical Chemists", AOAC, Washington, D.C., Thirteenth Editon, 1980.
6. P.J. Van Niekerk, "HPLC in Food Analysis", R. Macrae (Ed.), Academic Press, London, 1982, p. 187.
7. G.R. Skurray, Fd Chem., 1981, 7, 77.
8. A.T. Rhys Williams and W. Slavin, Chromatogr. Newsl., 1977, 5, 9.
9. I.D. Lumley and R.A. Wiggins, Analyst, 1981, 106, 1103.
10. V.A. Bognar, De LebensmittRdsch., 1981, 77, 431.
11. K. Dakshinamrti, and M.S. Chauhun, "Methods in Vitamin B6 Nutrition. Analysis and Status Assessment", J.E. Leklem and R.D. Reynolds (Eds), Plenum Press, New York, 1981, p.99.
12. J.T. Vanderslice, C.E. Maire, R.F. Doherty and G.R. Beecher, J. Agric. Fd Chem., 1980, 28, 1145.
13. K.L. Lim, R.W. Young and J.A. Driskell, J. Chromatogr., 1980, 188, 285.
14. J.F. Gregory, J. Fd Sci., 1980, 45, 84.

15.   J. F. Gregory and J.R. Kirk, "Methods in Vitamin B$_6$ Nutri-
      tion.    Analysis and Status Assessment", J.E. Leklem and
      R.D. Reynolds (Eds), Plenum Press, New York, 1981, p.149.
16.   J.T. Vanderslice,  C.E.  Maire  and  J.E.  Yakupkovic,
      J. Fd Sci., 1981, 46, 943.
17.   J.F. Gregory, J. Agric. Fd Chem., 1980, 28, 486.

# Fluorimetric Methods: Applications and Limitations

By W. F. Shipe

DEPARTMENT OF FOOD SCIENCE, CORNELL UNIVERSITY, 118 STOCKING HALL, ITHACA, N.Y. 14853, U.S.A.

### Introduction

Fluorescence results from absorption of light and subsequent emission of light by a molecule. The light emitted is of lower energy by an amount corresponding to that lost by vibrational relaxation. The ability of a molecule to fluoresce is dependent on its structure and its environment[1,2]. In some cases it is feasible to modify the structure by chemical means and thus alter fluorescence. In fact, many fluorimetric assays depend on the conversion of non-fluorescent molecules to fluorescent molecules or complexes by appropriate chemical treatment. Although structure determines the intrinsic fluorescent characteristics of molecules, environmental factors have a major impact on analytical results. The use of fluorimetry as an analytical tool has been the subject of investigations for about 150 years. Development of fluorimetric procedures was slow at first because of the lack of reliable instruments. In the past 25 years there have been significant improvements in the instrumentation and more rapid development of fluorescent techniques. Development of the Xenon arc lamp contributed to progress by providing an energy output continum throughout the ultraviolet and visible regions (230-700 nm). Most of the research to date has been focused on the use of fluorimetric methods for quantitative analysis. Recently, fluorimetry has also been used to study molecular structure and chemical interactions[3,4,5,6]. The general principles, limitations and applications of fluorimetric techniques to biological materials have been reviewed in detail by Udenfriend [7,8]. This review will deal with the fluorescent techniques that are applicable to dairy products research.

## Quantitative Analysis

Principles and limitations. Fluorimetry is a valuable quantitative tool because of its high sensitivity. It can be used to measure concentrations as low as $10^{-8}$ to $10^{-10}$ g/ml whereas colorimetric methods can rarely detect concentrations as low as $10^{-7}$ g/ml. The sensitivity of fluorimetric methods is dependent on the intensity of fluorescence. This intensity is equal to the intensity of the absorbed light multiplied by the quantum efficiency of fluorescence. This relationship is expressed mathematically in the following equation:

$$F = [I_O (1-10^{-\varepsilon cd}] [\phi]$$

where F = total fluorescence intensity quanta per second

$\quad$ $I_O$ = intensity of exciting light; quanta per second

$\quad$ c = concentration of solution

$\quad$ d = optical depth of solution

$\quad$ $\varepsilon$ = molecular extinction coefficient

$\quad$ $\phi$ = quantum efficiency (yield) of fluorescence

The quantum yield $\phi$ may be calculated as follows:

$$\phi = \frac{\text{number of quanta emitted}}{\text{number of quanta absorbed}}$$

The quantum yield may be affected by such factors as concentration, solvent, temperature, adsorption on container surface, viscosity and either intramolecular or intermolecular interactions.

At low concentrations fluorescence intensity is a linear function of concentrations. Fluorescence intensity increases as the concentration increases until it reaches a limiting maximum. After reaching the maximum, the apparent fluorescence decreases for right angle or end-on detectors whereas it remains unchanged for surface detectors. The effects of the different detector arrangements and concentration on the apparent fluorescence is shown in Figure 1. Fluorimeters for analyzing solutions normally have the right-angle geometry.

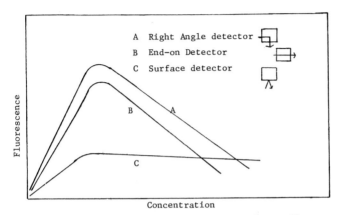

Figure 1. Effect of fluorophore concentration on fluorescence.

The non-linearity of fluorescence intensity at high concentrations may be due to either an "inner-filter effect" or "molecular interactions" or both. The inner filter effect can be caused by either absorption of fluorescence emission by a solution or absorption of a significant fraction of exciting radiation by some species in solution. It has been suggested that the inner filter effect could be avoided by using concentrations that do not exceed 5% absorption[1].

High concentrations of the fluorescent compound can also lead to the formation of a transient dimer composed of an excited molecule and a ground-state molecule[1]

$$A^* + A \longrightarrow (AA)^*$$

This dimeric species $(AA)^*$ has been called an excimer. This dimer will have a different electron orientation and longer emission wavelength than the monomer. Thus the fluorescence intensity at the "monomer's emission wavelength" will be reduced.

Excimer formation is a special type of intermolecular interaction that reduces or eliminates fluorescence. Quenching is the general term that is used to describe all interactions that suppress fluorescence. In general there are two types of fluorescence quenching: collisional and static. Collisional quenching is a diffusion-controlled interaction between an excited molecule and a quencher molecule. (The quencher molecule is in the ground state. It may be either the same type of molecule as the excited one or a different type.) Static quenching involves complex formation between a potentially fluorescent molecule in the ground state and a quencher. (The quencher in this case could be either the same type of molecule or a different one.) Concentration of a sample can affect both collisional and static quenching.

To determine if the concentration is too high the analyst can determine if fluorescence decreases on dilution. An increase in fluorescence on dilution would indicate that the initial concentration was too high (*i.e.* readings were being taken on the negative slope of curve A in Figure 1).

The nature and purity of the solvent can have significant effects on fluorimetric measurements [1,7,8,9]. In view of the high sensitivity of these measurements, traces of either fluorescing or quenching impurities can have marked effects. Some stopcock greases have been reported to contain fluorescent impurities. Impurities can be adsorbed on container surfaces so they should be thoroughly cleaned. One should exercise care in the choice of the cleansing agents since they may cause interference. For example, chromic acid may leave a film that absorbs ultraviolet light. This problem did not arise when nitric acid was used [7]. The polarity of the solvent can also affect fluorescence. Changing from polar to nonpolar solvents can produce large shifts in the frequency of emission. In some cases, the effect of solvent changes is due to hydrogen bonding. The selective effects of different solvents can be utilized in analyzing mixtures of fluorescing compounds. For example, the fluorescence of butylated hydroxy anisole (BHA) is quenched by a 80:20 mixture of ethanol and chloroform whereas propyl gallate (PG) fluorescence is not quenched by this solvent[9]. Therefore, the concentration of PG in a PG-BHA mixture could be measured without interference from BHA fluorescence. Obviously, the analyst should run appropriate solvent controls to make sure that the solvent is not interfering with the assay.

Fluorescence intensity usually decreases as temperature increases. At higher temperatures excitation energy may be dissipated more readily in other ways than by fluorescence emission. The magnitude of the effect depends on both the fluorescing compound and the solvent. For example, tryptophan is much more sensitive to temperature change than tyrosine or quinine[7]. Since most fluorimeters use high intensity lamps heating may occur in the instruments. If the sample being tested is particularly temperature sensitive it will be necessary to regulate the temperature of the sample holder. It has been reported[1] that viscosity affects fluorescence which is in addition to the temperature effect. Both temperature and viscosity could affect intermolecular interactions and the amount of dissolved gases. The fluorescence of a few oxygen sensitive compounds can be quenched by dissolved oxygen.

A number of fluorimetric procedures have been developed or adapted for assaying substances found in dairy products. The substances that have been assayed are listed, along with pertinent references, in Table 1. Multiple references are cited if more than one procedure is believed to be currently in use. There are fluorimetric methods for many other substances that have not yet been applied to dairy products. Many of these methods can be adapted for dairy product analysis with very little modification. In some cases the procedures must be modified to remove interfering substances or to compensate for their presence.

Table 1. Applications of Fluorimetric Analysis to Substances Found in
Dairy Products

| Substances | References | Substances | References |
|---|---|---|---|
| 1. Vitamin A | 10,11,12 | 12. Aflatoxin | 36 |
| 2. Thiamin | 13,14 | 13. Antibiotics | 37,38 |
| 3. Riboflavin | 14,15 | 14. Carrageenan | 39 |
| 4. Vitamin $B_6$ | 16,17 | 15. Cholesterol | 40 |
| 5. Folic Acid | 18 | 16. Histamine | 41-43 |
| 6. Vitamin C | 19-22 | 17. Pyruvate | 44 |
| 7. Selenium | 23,24 | 18. Sulfhydryl | |
| 8. Zinc | 25 | and disulfides | 45 |
| 9. Tryptophan | 26 | 19. Volatile thiols | 46 |
| 10.Available lysine | 27 | 20. Carboxyl esterase | |
| 11.Amines; amino acids | | activity | 47 |
| and peptides | 28-35 | 21. Microbial pigments | 48 |

In milk many of the potential interfering substances can be removed by conventional separation procedures such as centrifugation, filtration, solvent extraction and chromatographic techniques. In our laboratory, protein interference in the pyruvate test was removed by ultrafiltration. In view of the new ultrafiltration membrane technology, ultrafiltration will probably be used more frequently in the future for removal of interfering substances. Enzymatic treatments can also be used for removing interfering substances in some cases. For example, enzymatic digestion was used to eliminate protein "interference" in a carrageenan assay[39]. The carrageenan was adsorbed on the protein.

Most of the early quantitative assays were limited to compounds that fluoresce. Now many of the studies involve the preparation of fluorescing derivatives from non-fluorescing compounds. Sometimes compounds that fluoresce are derivatized in order to shift the emission maximum. This technique can be used to avoid overlapping emission spectra in samples with mixtures of fluorescing compounds.

Extensive research has been conducted on the various fluorescent derivatives of amines and amino acids. The p-dimethylaminonaphthialene-sulfonyl (dansyl) reagent was used to prepare derivatives of amines, amino acids, peptides and proteins[28]. A large excess of dansyl chloride is required for the preparation which may lead to disubstitution on the nitrogen. The excess of reagent and its degradation products must be removed before the derivatives can be separated chromatographically. Ninhydrin derivatives are reported[29,30] to be superior and suitable for automated method. Ninhydrin is not fluorescent so excess reagent does not interfere. However, the preparation of ninhydrin derivatives requires a heating step. The fluorescent ninhydrin method is not affected by ammonia whereas both the dansyl and colorimtric ninhydrin procedures are. The ninhydrin studies led to the development of a new derivatizing agent, namely 4-phenylspiro[furan-2(3H),1'-phthalan]-3,3'dione (fluorescamine)[31]. Fluorescamine reacts very rapidly and can be used in either aqueous solutions or organic solvents. A fluorescent chromophore formed by the reaction of amino acids with o-phthaldialdehyde (OPA) in the presence of thiols has been studied by several investigators[32-35,41]. The fluorescent yield of OPA derivatives of amino acids is reported to be about 10 times greater than for fluorescamine derivatives. The fluorecence yield of OPA derivatives of peptides is much lower than it is for amino acids. Because

of  this difference and the speed of reaction the OPA method has been  used
for peptidase activity measurements[49,50].

     Although  the OPA reagent is quite versatile it has limitations.   For
example,  it does not react with proline or hydroxyproline and gives a very
low fluorescence with cysteine.   The fluorescence of disubstituted  lysine
is   lower  than  monosubstituted  lysine.    The  quantum  yield  of   the
disubstituted  lysine increased in the presence of sodium  dodecyl  sulfate
(SDS) according to Chen et al[33].  They postulated that quenching was due to
interaction  between  the  two  isoindole groups  on  lysine  and  that  SDS
suppressed  this interaction.   Porter et al[50] suggested that the effect of
SDS  on the fluorescence of OPA derivatives was dependent on  the  distance
between interacting isoindole or indole group.

     The  OPA reagent has also been used to prepare amino acid  derivatives
for reversed-phase HPLC[34,35].  By precolumn derivatization, the polar amino
acids are converted to hydrophobic compounds that can be rapidly  separated
on  a reversed phase column.   The amino acids can also be separated on  an
ion  exchange  column  followed  by  post-column  derivatization.   If  the
derivatives are stable, precolumn derivatization may be preferable to post-
column  derivatization.   The  choice will depend on the relative  ease  of
separation of the derivatives versus the underivatized compounds.

     In  some  cases  derivatizing  agents can be  used  to  trap  volatile
compounds.   A  thiol  reagent,  N-4-(7-diethylamino-4-methylcoumarin-3-yl)
phenyl-maleimide  (CPM) was used to trap volatile thiols from milk that had
been  exposed  to  light[41].  The CPM was dissolved in  butanol  which  was
immersed  in a dry ice-acetone bath (-70°C).   Nitrogen was bubbled through
the milk into the trapping solution where the CPM reacted with the volatile
thiols to form  non-volatile fluorescent adducts.

### Qualitative Measurements

     Fluorimetric  techniques  have been employed by dairy  scientists  and health officials to detect microbial contamination of dairy products. These techniques are useful in detecting organisms that produce fluorescent pigments. The concentration of fluorescent pigments can provide an index of bacterial count in samples contaminated with a pigment producing bacterial species such as <u>Pseudomonas fluorescens</u>[48]. The presence of fluorecent pigments can also help to identify the types of microbial contamination. Fluorescence dyes are used for staining bacterial and somatic cells in milk.

Dairy scientists need to explore the use of fluorimetric techniques for studying molecular and micellar structure. Considerable progress has been made in elucidating the structure of proteins, membranes and micelles of other biological materials[5-8]. Creamer <u>et al.</u>[51] have used two fluorescent probes (<u>i.e.</u> <u>cis</u>-parinaric acid and 1,8-anilinonaphthalene sulfonate) to study the surface hydrophobicity of casein. Research of this type should also be useful in elucidating the structure of the fat globule membrane and the hydrophobic interactions between proteins and flavor.

<h2 style="text-align:center">References</h2>

1.  D.M. Hercules, ed. "Fluorescence and Phosphorescence Analysis", Interscience Publishers, New York 1966.

2.  G.G.Guilbault, ed., "Fluorescence", Marcel Dekker, Inc, New York, 1967.

3.  R.F. Chen and H. Edelhoch, ed., "Biochemical Fluorescence-Concepts", Marcel Dekker, Inc, NY, 1975, Vol 1.

4.  R.F. Chen and H. Edelhock, eds. "Biochemical Fluorescence-Concepts", Marcel Dekker, Inc, NY, 1976, Vol 2.

5.  E.L. Wehry, ed. "Modern Fluorescence Spectroscopy", Plenum Press, New York and London, 1976, Vol 1.

6.  E.L. Wehry, ed. "Modern Fluorescence Spectroscopy", Plenum Press, New York and London, 1976, Vol 2.

7.  S. Udenfriend, "Fluorescence Assay in Biology and Medicine", Academic

Press, London 1962 Vol. I.

8.   S. Udenfriend, "Fluorescence Assay in Biology and Medicine", Academic Press, London, 1969 Vol. II.

9.   R.J. Hurtubise, Anal. Chem., 1976, 48, 2092.

10.  J.N. Thompson, P. Erdody, W.B. Maxwell & S.K. Murray. J. Dairy Sci., 1972, 55, 1077.

11.  J.N. Thompson & R. Madere, J. Ass. Off. Analyt. Chem., 1978, 61, 1370.

12.  G.F. Senyk, J.F. Gregory & W.F. Shipe, J. Dairy Sci., 1975, 58, 558.

13.  P.W. Defibausch, J.S. Smith & C.E. Weeks, J. Ass. Off. Analyt. Chem., 1977, 60, 522.

14.  W.E. Dumber & K.E. Stevenson, J. Ass. Off. Analyt. Chem., 1979, 61, 642.

15.  I. Rashid and D. Potts, J. Food Sci., 1980, 45, 744.

16.  V. Friedlerova & J. Davidek, Z. Lebensmitteluntersuchung und Forsch., 1974, 155, 277.

17.  J.F. Gregory & J.R. Kirk, J. Food Sci., 1977, 42, 1073.

18.  J.F. Gregory, D.B. Manley & B.P.F Day, Federation Proceedings, 1983, 42, 667.

19.  J.R. Kirk & N. Tins, J. Food Sci., 1975, 40, 463.

20.  R.B. Roy, A. Conetta & J. Salpeter, J. Ass. Off. Analyt. Chem., 1976, 59, 1244.

21.  D.C. Eshers, R.H. Potter & J.C. Heroff, J. Ass. Off. Analyt. Chem., 1977, 60, 126.

22.  D.L. Dunnuri, J.D. Riese, R. Byron & M. Sessers, J. Ass. Off. Analyt. Chem., 1979, 62, 648.

23.  N.D. Michie, E.J. Dixon & N.G. Burton, J. Ass. Off. Analyt. Chem., 1978, 61, 48.

24. J.H. Watkinson, Anal. Chim. Acta, 1979, 105, 319.

25. A. Moreno, M. Silva, D. Perez Dendito & M. Valcarcel, Analyst, 1983, 108, 85.

26. H.Steinhart, Z. Tierphysiol. Tierenahr. Futtermittelk., 1978, 41, 48.

27. C. Goodno, H.E. Swaisgood & G.L. Catignani, Anal. Biochem., 1981, 115, 203.

28. W.R. Gray & B.S. Hartley, Biochem. J., 1963, 89, 379.

29. K. Samejima, W. Dairmon & L. Udenfriend, Anal. Biochem., 1971, 42, 222.

30. K. Samejima, W. Dairman & L. Udenfriend, Anal. Biochem., 1971, 42, 237.

31. S. Udenfriend, S. Stein,P. Bohler, W. Dairman, W. Leimgruber & M. Weigele, Science, 1972, 178, 871.

32. S. Taylor & A.L. Tappel, Anal. Biochem., 1973, 56, 140.

33. R.F. Chen, C. Scott & E. Trepman, Biochim. Biophys. Acta, 1979, 576, 440.

34. D.W Hill, F.H. Walters, T.D. Wilson & J.D. Stuart, Anal. Chem., 1979, 51, 1338.

35. T.A. Kan & W.F. Shipe, J. Food Sci., 1981, 47, 338.

36. R. Gauch, U. Leuenberger & E. Baumgartner, J. Chromatography, 1979, 178, 543.

37. H. Poiser & C. Schlatter, Analyst, 1976, 101, 808.

38. J. Hamann, W. Hesschen, A. Tolle, & G. Hahn, Milchwissenschaft, 1975, 30, 139.

39. D. Murray & R.B. Cundall, Analyst, 1981, 106, 335.

40. T. Badzio, U. Winnicka & A. Soluki, Chemia Analtyzna, 1975, 20, 701.

41. S.L. Taylor & E.R. Lieber, J. Food Sci., 1977, 42, 1584.

42. T.L. Chambers, & W.F. Staruszkiewicz, Jr., J. Ass. Off. Analy. Chem.,

1978, <u>61</u>, 1092.

43. P.H. Larsen, C. Mikkelsen & E. Waasher Nielsen, <u>Dansk Veterinaertidsskrift</u>, 1981, <u>64</u>, 878.

44. T.M. Cogan, <u>Irish J. Food Sci. & Technol.</u>, 1977, <u>1</u>, 143.

45. S.D. Senter, W.K. Stone and W. C. Thomas, <u>J. Dairy Sci.</u>, 1973, <u>56</u>, 1331.

46. W.J. Liu, Master's Thesis, Cornell University, Ithaca, NY 1983.

47. H.C. Deeth, XX International Dairy Congress, 1978, <u>E</u>, 364.

48. W.F. Shipe and H.Y. Hsu, <u>J. Dairy Sci.</u>, 1982, <u>45</u>,Suppl. 1, 45.

49. S. Taylor and A.L. Tappel, <u>Anal. Biochem.</u>, 1973, <u>56</u>, 140.

50. D.H. Porter, H.E. Swaisgood and G.L. Catignani, <u>Anal. Biochem.</u>, 1982, <u>123</u>, 41.

51. L.K. Creamer, H.F. Zoerb, N.F. Olson and T. Richardson, <u>J. Dairy Sci.</u>, 1982, <u>65</u>, 902.

# Some Limitations of *in vitro* Methods for the Assay of B Complex Vitamins in Foodstuffs

By J. E. Ford

NATIONAL INSTITUTE FOR RESEARCH IN DAIRYING, SHINFIELD, READING, BERKSHIRE
RG2 9AT, U.K.

We are at a stage of rapid development of chromatographic and
other physico-chemical methods of vitamin assay.  HPLC has been
applied successfully in the measurement of all the fat-soluble
and several water-soluble vitamins in foods and tissues, and some
of the older established procedures tend to look somewhat old-
fashioned.  Certainly microbiological methods for the assay of B-
complex vitamins will come under increasing challenge.  But we
must take care lest greater speed and precision in measurement,
and the allure of expensive and sophisticated hardware, encourage
uncritical acceptance of the results.

In measuring vitamins in a foodstuff our aim is to predict
accurately the  effective  or 'biologically available' content
of vitamin in that foodstuff as it is used in the nutrition of man
or beast.  For assays with chicks and other laboratory animals we
offer the test sample as an ingredient in the diet.  But for in
vitro tests we must first extract the vitamin and then sometimes
'clean up' the extract.  This need for extraction is the first of
our difficulties.  Formidable problems may then arise related to
the specificity of the test procedures, and interpretation of the
results in terms of biological activity for higher animals.  Each
of the B vitamins occurs in nature in a multiplicity of chemically
distinct though closely related forms, which may be 'free' or −
more commonly − firmly conjugated with protein.

The resulting problems of measurement are well exemplified in the
assay of folic acid and vitamin $B_{12}$ in milk.

The older nutrition text books and tables of food composition
show cow's milk as being poor in folate.  This low rating
reflects the one-time popularity of Streptococcus faecalis as test
organism in the microbiological assays.  But the predominant form
of folate in fresh milk, as in eggs and liver and other animal

tissues, is methyltetrahydrofolate, which is inactive for
Strep.faecalis.  Analysts now generally use Lactobacillus casei
for their microbiological tests;  it responds to the methyl
folate, and in consequence milk appears greatly improved as a
source of folate.  And indeed there is no doubt that the lower
values obtained with Strep.faecalis were incorrect.

As we see, the choice of test microorganism may determine the
results, in this as in assays for several other B-vitamins.  The
results may also be influenced by the form of the vitamin
selected as the reference standard.  In the folate assay, folic
acid is generally used.  It is the parent compound of a numerous
tribe of derivatives that occur in natural materials mainly as
polyglutamyl forms of tetrahydrofolate (THF).  To use it as the
standard against which to compare - for example - 5-Me THF (the
predominant form of folate in food extracts after the required
treatment with γ-glutamyl-carboxypeptidase) has been reported to
give misleading results;  Phillips & Wright[1,2] maintain that many
folate values as given in current Tables of Food Composition are
erroneously low.  A major problem in their Lb.casei assays was
that the relative growth-promoting activities of FA and 5-Me THF
varied with pH of the assay medium.  At pH 6.8 the response to
Me THF was lower than that to FA and there was marked non-
linearity in the assays.  But at pH 6.1 the two congeners had the
same activity and there was no evidence of drift.  An analogous
problem has been reported for the competitive binding radioassay
for folate, in that the relative affinities of FA and 5-Me THF for
folate-binding protein varied with pH[3].  And as with the
microbiological assay, it was possible to specify test conditions
such that the two vitamers reacted equally.

Besides the choice of test organism, other features of the
microbiological assay may influence the results.  Lb.casei
responds little if at all to the natural polyglutamyl forms of
folate, which we assume (from inconclusive evidence) to be fully
available to higher animals.  So these conjugates must be
converted to mono- or diglutamates under the action of γ-glutamyl
carboxypeptidase, or 'deconjugase enzyme'.  Various forms of this
enzyme occur in nature, those present in chick pancreas and pig
kidney having been most widely used in folate assay.  But no
ideal conjugase treatment applicable to all types of test sample

has yet been standardized, and it is generally necessary to carry out a preliminary study on extraction with each new type of test material.

There are other complications in the assay process, besides uncertainties concerning the efficiency of the deconjugation procedures and the nutritional equivalence of different natural forms of folate. For example, during storage of milk and presumably of other foodstuffs, 5-Me THF oxidizes readily to 5-Me 5:6-dihydrofolate which, under acid conditions as in the stomach, is rapidly converted to a biologically inactive form[4,5]. But the dihydrofolate is fully active for Lb.casei in presence of the ascorbic acid that is added to the test samples during extraction. So we have a situation in which the natural form of the vitamin is regenerated as an artefact in the process of measurement.

There is still a need for comparative biological and microbiological testing to establish assay procedures capable of predicting accurately the biological activity of folate as it occurs in a wider variety of foodstuffs. This empirical procedure for checking the in vitro assay results is adequate for the practical purposes of the nutritionist. More detailed analysis of the folate composition, as by HPLC, may reveal a bewildering assortment of folate-active compounds but give little further useful information concerning the nutritional value of the foodstuffs as sources of folate. Fig.1 illustrates this point, showing the apparent multiplicity of folate congeners and derivatives in pasteurized milk, as revealed by chromatography in Sephadex G10.

Problems concerning extraction, and specificity of the assay procedures, are of course not confined to the measurement of folate. They are of common occurrence in vitamin assay and should be well understood by the experienced analyst. There remains however a need for critical understanding of the relevance of our assay procedures to the purposes for which the tests are carried out. Thus, for example, amino acid analysis is an indispensable aid in the evaluation of food proteins, and is widely employed in quality control in the food industry. Yet among animal feed protein concentrates of similar designation (e.g. fish meals) we may find large differences in nutritional quality but only very

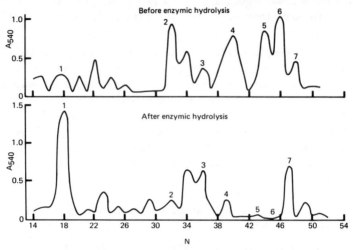

Fig.1   Separation of folates from pasteurized whole milk on Sephadex G-10 as evaluated by the microbiological test with *L. casei*. N = Number of fractions each of 3.3 ml. Peaks: 1 = $N^5$-methyltetrahydrofolic acid; 2, 4, 5 and 6 = unidentified polyglutamates; 3 = $N^{10}$-formyldihydrofolic acid; 7 = folic acid. *From Kas & Cerna (6).*

small differences in amino acid composition. The differences in protein quality reflect differences in digestibility that are no longer apparent in the acid hydrolysates prepared for amino acid analysis. In connexion with this particular problem, it has been said that "the advent of more accurate methods of total amino acid analysis offers only a more refined means of obtaining the wrong answer, though with greater precision than hitherto". The same criticism may sometimes apply in our vitamin assays: we must keep in mind that they give only the crudest data on composition and they can be misleading.

Paediatricians and baby milk manufacturers are increasingly aware that even the most sophisticated formula milks are a poor substitute for the maternal milk, which is not merely an optimal assemblage of nutrients, but rather a finely integrated life support system. Part of the explanation for the superiority of breast milk has to do with the biological availability of the nutrients. Thus with folate and vitamin $B_{12}$ and probably other vitamins, and with iron and zinc, the form in which they are presented may influence their value to the baby. Assays of breast milk tell us that it contains about 50 µg folate/ℓ, and most formulas are supplemented with folic acid to ensure that they contain at least this amount. But folic acid is not a natural form of folate in milk and indeed it has no biological activity until it has undergone enzymic reduction. Barford & Pheasant[7] point out that folic acid is a poor substrate for mammalian dihydrofolate reductase, and it inhibits dihydropteridine reductase, a key enzyme in neurotransmitter biosynthesis. They suggested that the practice of giving folic acid to expectant mothers should be reviewed, and that meanwhile it would be prudent to give a reduced folate instead. Perhaps even more so with artificially reared premature and weakly babies, we should question whether this precursor form of the vitamin is a proper substitute for the natural milk folate.

Even if we were to add this expensive and labile natural folate to our milk replacement formulas, we would still have problems. There is in fact no free folate in fresh milk, human or bovine. The vitamin is locked away in a specific folate-binding protein, and so protected against uptake by intestinal bacteria. Yet at the same time it is readily taken up from the intestine - more so apparently than is free folate[8]. The same situation obtains with vitamin $B_{12}$ in human milk. Like folate it occurs in a protein-bound form that is unavailable to microorganisms. Indeed this binder first came to notice as a bacterial growth inhibitor in the assay of vitamin $B_{12}$ in sow's milk[9]. And this bound form of the vitamin in sow's milk is much more readily absorbed from the intestine of the neonatal piglet than is the free vitamin (Dr N. Trugo, personal communication).

Quite probably a similar story could be told concerning the biological availability of other minor nutrients in milk. Clearly compositional analysis of milk does not take us all the way along the road to an understanding of its special nutritional importance for the neonate. We need to know much more about the subtle mechanisms whereby the constituents of the maternal milk are so efficiently dedicated to the benefit of the recipient infant.

## References

[1] D.R. Phillips and A.J.A. Wright, Br. J. Nutr., 1982, 47, 183.

[2] D.R. Phillips and A.J.A. Wright, Br. J. Nutr., 1983, 49, 181.

[3] J.K. Givas and S. Gutcho, Clin. Chem.,1975, 21, 427.

[4] J.E. Ford, K.J. Scott and J.A. Blair, Proc. Intl Dairy Congr., 1978, 20 Paris E1069.

[5] K. Ratanasthien, J.A. Blair, R.J. Leeming, W.T. Cooke and V. Melikian, J. Clin. Path., 1977, 30, 438.

[6] J. Kas and J. Cerna, J. Chromatogr., 1976, 124, 53.

[7] P.A. Barford and A.E. Pheasant, Br. Med. J., 1981, 282, 1793.

[8] N. Colman, N. Hettiarachchy and V. Herbert, Science, 1981, 211, 1427.

[9] M.E. Gregory, J.E. Ford and S.K. Kon, Biochem. J., 1952, 51, xxix.

# Automated Methods for Potentiometric Determinations in Dairy Products, Especially Sodium Chloride in Cheese

By M. Collomb and G. Steiger*

FEDERAL DAIRY RESEARCH INSTITUTE, CH-3097 LIEBEFELD-BERN, SWITZERLAND

Introduction

Automation of potentiometric titrations can offer many advantages
to routine analytical laboratories. Firstly it greatly enhances
the working capacity of the laboratory due to the utilization of
an automatic sampler. Secondly it improves the accuracy of the
results due to the use of an automatic titrator which gives the real
endpoint for each titration. Finally, the correctness of the calculations of the result is fully warranted.

Description of the system *

The system (Figure 1, Plate 1) consists of a titrator ("TITROPRO-
CESSOR") connected on one side with two automatic burettes and a
balance with its terminal and printer and on the other side with
the sampler control unit.

By pressing a key at the balance terminal the weight of the sample
is introduced into the TITROPROCESSOR and printed for control. The
samples are then placed in the sampler (32 places). If necessary,
the samples may also be numbered.

---

* Balance part of the system:    METTLER Instrument Ltd

                                         CH-8606 Greifensee (Switzerland)

    Titration part of the system:    METROHM Ltd, CH-9100 Herisau (Switzerland)

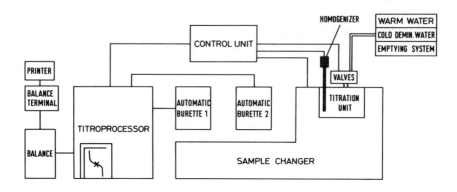

Figure 1 : Block diagram of the titration system

Plate 1: General view of the system

The following operations (see a, b, c below) are performed automatically by the control unit which may contain up to 8 different
programmes, thus allowing the user to vary to his liking the use
and operation time of a homogenizer, of a stirrer and of filling,
rinsing or emptying valves etc. During the determination of sodium chloride in cheese, the sequence of operations is as described below (Figure 2, Plate 2):

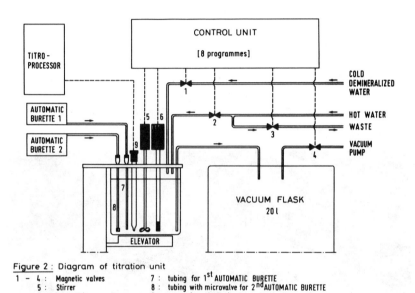

Figure 2 : Diagram of titration unit

| | | | |
|---|---|---|---|
| 1 – 4 : | Magnetic valves | 7 : | tubing for 1st AUTOMATIC BURETTE |
| 5 : | Stirrer | 8 : | tubing with microvalve for 2nd AUTOMATIC BURETTE |
| 6 : | Homogenizer | 9 : | measuring electrode |

a) Homogenization_of_the_sample

  - The sample is raised under the titration head with a
    mobile arm (elevator).

  - The valve for demineralized water ① opens and approxi-
    mately 100 ml of demineralized water are added to the
    cheese. The water comes from a source regulated at a
    pressure of 0.5 bar.

- The stirrer ⑤ is switched on and remains so until
  rinsing has terminated.

b) <u>Titration</u>

- 2 ml of 4 M nitric acid are added by the first
  automatic burrette ⑦ under continuous stirring

- Simultaneously the mV-scale is printed on the regis-
  tration paper of the printer of the TITROPROCESSOR
  (plate 1).

- Titration with 0.1 M silver nitrate is carried out
  by the other automatic burette ⑧ . The extremity
  of the point of this burette contains an antidif-
  fusion tip.

Plate 2: View on titration head and sampler

- The titration is conducted by the TITROPROCESSOR
which also records the titration curve; the end point
is indicated on the titration curve by an asterisk.
The visualization of the titration curve allows one
to control the proper operation of the electrode
and the correctness of the titration.

- The volume of silver nitrate and the result expressed
in g NaCl per 100 g cheese are printed on the regis-
tration paper.

## c) Emptying and rinsing

- As soon as the titration is completed, valves ③
opens (valve ② remains closed) to let warm water
heat the system in order to have water hot enough
for the subsequent rinsings of the titration head.

- Valve ③ closes and valve ④ opens to empty the
titration vessel. This is done by a vacuum generated
in a flask of 20 l by a water pump. This system proved
to be the most reliable. The emptying time has to be
regulated to restore atmospheric pressure in the vacuum
flask.

- Valve ④ closes and valve ② opens so that approximately 100 ml of warm water run into the titration
vessel; the wash-water is then evacuated as described
above. This step is repeated with the homogenizer
switched on for 15 sec. After the homogenizer stops,
the vessel is emptied and lowered. The next cycle can
then begin.

## Remarks

All operations mentioned above are contained in only one of the
8 possible programmes of the control unit.

As soon as the first sample has been weighed, the titration can begin and the other samples can be weighed after titration has started. One titration, i.e. one cycle, takes about 6 minutes.

The titration system works in the so called "dynamic mode", i.e. when it nears the end-point the volume of titrating solution added becomes smaller and smaller.

Applications
_____

Determination of sodium chloride in cheese (1) :

|         | Swiss cheese | extrahard cheese |
|---------|:------------:|:----------------:|
| N       | 10           | 10               |
| $\bar{x}$ | 0.451      | 2.004            |
| $s_x$   | 0.011        | 0.007            |

Determination of the acidity of milk (2) and cream (3) in $^{\circ}SH$ :

|         | Titration to real equivalent point (pH 8.5 - 8.6) | Titration to pH 8.3 | Titration with phenol-phthalein |
|---------|:---:|:---:|:---:|
| N       | 5    | 5    | 5    |
| $\bar{x}$ | 6.82 | 6.53 | 7.7  |
| $s_x$   | 0.08 | 0.07 | 0.1  |

The system described is very versatile and can be used for many other applications such as the determination of chloride in milk (4), sodium chloride in butter (5), vitamin C in milk (6), acidity of butterfat (7), iodine value in butter fat (8) etc.

The system has been conceived in such a way as to work also without the sampler, which is advantageous when dealing with short series.

## Conclusions

Thanks to its flexibility and versatility the system described can be easily adapted to the specific needs of each laboratory. Intensive utilization in our laboratory proved that the system is highly reliable and that it brings a welcome rationalization to laboratories.

## Literature

(1)  International Standard IDF 88, 1979
(2)  Schweiz. Lebensmittelbuch, 5. Aufl., 2. Bd., Methode 1/13
(3)  Schweiz. Lebensmittelbuch, 5. Aufl., 2. Bd., Methode 3A/08
(4)  International Standard IDF 88, 1979, adapted for milk
(5)  International Standard IDF 12A, 1969 *
(6)  Official Methods of the AOAC, 13rd ed., 1980, method 43.056
(7)  International Standard IDF 6A, 1969 *
(8)  International Standard IDF 8, 1959 *

* with potentiometric end point determination

# Detection of Fractionated Butterfat by Crystallization of the Higher Melting Saturated Triglycerides

By B. G. Muuse* and H. J. van der Kamp
RIKILT, BORNSESTEEG 45, 6708 PD, WAGENINGEN, THE NETHERLANDS

Abstract

For detecting fractionated hard fractions of butterfat, the analysis of fatty acid profile, fatty acids in two position, cholesterol amount or triglyceride composition gives only slight indications because of variability both in product and analysis. A
reliable detection method is based on the crystallization of the
increased amount of saturated higher triglycerides which occur in
fractioned butterfat. The crystallization is done with hexane at
12.5°C. Normal butterfat from either summer or winter period does
not produce more than 0.5% crystals, whilst hard fractions of
butterfat will give 5-15% crystals depending upon the degree of
fractionation. Obviously, when over 0.5% crystals is found, the
occurrence of fractionated butterfat is proved.
Based on the same principle, the detection of soft fractions of
butterfat may be possible as well.

Introduction
In the Netherlands the declaration of butter is strictly regulated. The use of fractionated butterfat was reason for adjusting
these labelling regulations and for making a distinction between
fractionated and non-fractionated butterfats.
For controlling the ingredients labelling in the laboratory a
method had to be developed. For developing such a method the influence of the fractionation on the chemical composition of the
butterfat was studied.

* author for answering queries.

Butterfat can be fractionated by physical or chemical means. In the Netherlands butterfat is fractionated only by physical techniques.

Other techniques are not allowed nowadays. By this fractionation the butterfat is melted, crystallized by slowly cooling down and final filtering of the crystals on a vacuum filter. The obtained mass of crystals contains still about 70% enclosed liquid oil [1]. Beside this hard fraction a soft liquid-like fraction of butterfat is obtained. Especially the hard fractions are used by confectioners for preparation of puff pastry products (croissants, crusty cakes etc.)[2]. Hard fractions of butterfat are increasingly used in the Netherlands confectionery and in chocolate. The soft fractions of butterfat are almost exclusively applied in the production of ice-cream.

The chemical characteristics of butterfat are somewhat influenced by the fractionation, but they hardly cross the limits which normally characterise butterfat[3,4].

The ratio of C18:0/C18:1 fatty acids is the most striking change and mentioned by a lot of investigators[3,5,6,8,9,10,11]. The long-chain saturated fatty acids occur in increasing amounts in the hard fractions of butterfat; combined with low butyric acid and caproic acid contents this will give some indication of the presence of fractionated butterfat (Table 1).

The preference of fatty acids with carbon number 12 and higher for the center position of the triglyceride was not influenced by fractionation (Table 2).

The cholesterol from the butterfat is equally divided over the hard and soft fraction (Table 3).

Table 1. Fatty acid profile

|       | Hard | Original | Soft |
|-------|------|----------|------|
| C 4:0  | 3.3  | 4.0      | 4.2  |
| C 6:0  | 1.8  | 2.2      | 2.3  |
| C 8:0  | 1.1  | 1.3      | 1.3  |
| C10:0 | 2.4  | 2.7      | 2.8  |
| C12:0 | 3.3  | 3.3      | 3.4  |
| C14:0 | 10.4 | 10.0     | 9.9  |
| C16:0 | 27.0 | 24.5     | 23.9 |
| C18:0 | 14.5 | 12.1     | 11.2 |
| C18:1 | 24.5 | 27.5     | 28.0 |

Table 2. Percentage of preference for center position

|       | Hard | Original | Soft |
|-------|------|----------|------|
| C12:0 | 48   | 49       | 51   |
| C14:0 | 56   | 54       | 56   |
| C16:0 | 41   | 39       | 39   |
| C18:0 | 17   | 19       | 18   |
| C18:1 | 26   | 25       | 24   |

Table 3. Cholesterol percentage

| Hard  | Original | Soft  |
|-------|----------|-------|
| 0.231 | 0.278    | 0.291 |

Also this difference is too small for detecting hard fractionated butterfat in mixtures.

In the triglyceride composition an increase in saturated triglycerides with high carbon number C46-54 was found[6,7,8,10].

Parodi[12,14,15] analyzed the hard fractions on silver nitrate impregnated thin layer chromatography and notably found an increase of SSS, $SSM_{trans}$ and $SSM_{cis}$ triglyceride groups (S = saturated, M = monoene unsaturated fatty acid).

The classical characteristics: refraction index, density and iodine value are hardly changed by fractionation[6,7,8].

Especially the amount of solids in the hard fraction is increased and is a possibility for control purpose[7,8,10]. Some methods for determining the amount of solids are the determination of carotenes in the fat and the liquid part of the fat[6,17], the Nuclear Magn. Resonance determination of crystalline material[8,17] and the Differential Thermal Analysis[11,17].

The method we developed and describe here determines the amount of solids by crystallization in hexane at 12.5°C. At this temperature the non-fractioned butterfat from either summer or winter will not yield more than 0.5% of crystals. More crystals indicate a raised amount of higher-melting triglycerides indicating the presence of fractionated butterfat.

As this method is based on the composition of the triglycerides, the triglyceride composition is described too.

Material and methods

Protein free anhydrous butterfat. As material for studying the technique of analysis one bulk of fractionated butterfat was used. For studying the product characteristics several charges of butterfat were obtained consisting of: hard and soft fractions of butterfats along with the original butterfat belonging thereto and non-fractionated butterfat from summer and winter. The fractionated butterfats came also from other countries than the Netherlands.

For the evidence of summer and winter butterfat the ratio of the fatty acids C18:1/C16:0 and C18:0/C14:0 was determined.

For the summer season we found for the Dutch butterfat mean ratio's of ca. 1.1 both and for the winter season of ca. 0.8 both. (fig. 1)[13,18].

Triglyceride profile.
The triglycerides were
analyzed with capillary
gas chromatography on a
Varian 3700 gas chroma-
tograph fitted with a
fused silica column with
chemical bound phase of
CP Sil 5. about 10 me-
ter long, with internal
diameter of 0.23 mm.
Operating conditions:
Splitmode 1:20,
Carrier gas: Helium.
Flame Ionization Detector.
Data and integration
system SP 4000 of Spectra
Physics.

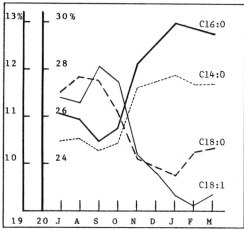

fig. 1 Seasonal variance in fatty
acid profile of butterfats

Injection volume 1μl of a 2% solution of fat in hexane.
Temperature of the oven is programmed from 260°C to 360°C with
4°C per minute.

Determination of solids by crystallization in hexane at 12.5°C.
Solve 25.0 g of melted fat in a conical flask with 50 ml of hexane. Put the conical flask in a waterbath cooled down till 12.5°C
± 0.2°C and stir magnetically till the first crystals are visible. Stop stirring and let crystallize for 30 minutes.
Note: Stop stirring if within 5 minutes no crystals are formed
and leave the solution for 30 min.
Filter off the crystals in an edgefolded filterpaper, flat pressed in a Büchner funnel, under
slight vacuum (fig 2). Quantitatively transfer
the crystals in the conical flask on the filter
by rinsing with the obtained filtrate.
Dissolve the isolated mass of crystals with
the filterpaper in 25 ml of warm hexane.

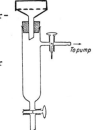

fig.2   Suction unit

Take the filterpaper out of the solution and wash it with 25 ml
of warm hexane. Collect the two 25 ml-fractions and crystallize
the fat again under the same conditions.
Collect the crystals similarly and finally wash the crystals on
the filterpaper with 50 ml of cold hexane (4°C).
Determine the mass of crystals by soxhlet extraction with pentane
and dry the fat during 1 h at 102°C.
Calculate the percentage of crystal mass from the original sample.


Results and discussion


Triglyceride study. The capillary chromatographic analysis of the
triglycerides gave a separation of the triglycerides by carbon
number and saturation degree. Traitler et al.[16] described this separation of triglycerides from milkfat. They identified the sub-
divided triglycerides with carbon number 54 as OOO, SOO, SSO and
SSS respectively, in successive retention time (S = Saturated
fatty acid, O = oleic acid).
The positional isomers were not separated.
In our study the different fractions of butterfat (hard, original
and soft) also obviously differ in triglyceride profile. (Table 4,
fig. 3 and chromatograms fig. 4).
Nevertheless this profile shows too small differences in fractions to be useful as a method for the detection of fractioned
butterfat in mixtures with normal butterfat.

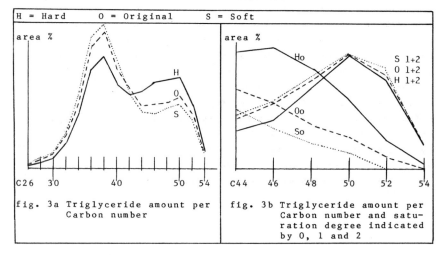

H = Hard        O = Original      S = Soft

fig. 3a Triglyceride amount per
        Carbon number

fig. 3b Triglyceride amount per
        Carbon number and satu-
        ration degree indicated
        by 0, 1 and 2

Table 4. Triglyceride composition in % area/area 0,1,2 and 3 corresponds with the number of double bonds in the T.G.

| C nr. | | Hard | Orig. | Soft |
|---|---|---|---|---|
| 26 | | 0.3 | 0.3 | 0.4 |
| 28 | | 0.6 | 0.7 | 0.9 |
| 30 | | 1.2 | 1.5 | 1.6 |
| 32 | | 2.5 | 3.2 | 3.5 |
| 34 | | 5.6 | 6.9 | 7.5 |
| 36 | | 9.9 | 12.3 | 13.3 |
| 38 | | 11.3 | 13.7 | 14.9 |
| 40 | | 8.5 | 10.0 | 10.7 |
| 42 | | 7.4 | 7.3 | 7.2 |
| 44 | 2+1 | 1.9 | 2.5 | 2.7 |
| | 0 | 5.9 | 4.0 | 3.2 |
| 46 | 2+1 | 2.5 | 3.2 | 3.4 |
| | 0 | 6.2 | 3.4 | 2.1 |
| 48 | 2+1 | 4.1 | 4.4 | 4.6 |
| | 0 | 5.0 | 2.3 | 1.2 |
| 50 | 2 | 1.1 | 1.3 | 1.2 |
| | 1 | 4.7 | 4.6 | 5.1 |
| | 0 | 3.5 | 1.6 | |
| 52 | 2 | 2.1 | 2.6 | 2.8 |
| | 1 | 2.5 | 2.2 | 2.3 |
| | 0 | 1.4 | 0.4 | 0.0 |
| 54 | 3 | 0.4 | 0.5 | 0.6 |
| | 2 | 0.6 | 0.8 | 0.8 |
| | 1 | 0.4 | 0.5 | 0.4 |
| | 0 | 0.3 | 0.0 | 0.0 |

The identification of the TG peaks is illustrated by: where: 0 = SSS, 1 = SSM, 2 = SMM, 3 = MMM

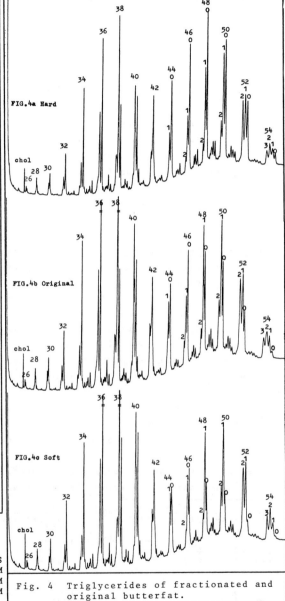

FIG.4a Hard

FIG.4b Original

FIG.4c Soft

Fig. 4　Triglycerides of fractionated and original butterfat.

Figure 3a shows that the triglycerides of C44-54 are enriched at the expense of the lower triglycerides in case of hard fractions. In figure 3b the differences in degree of saturation are outlined. It shows that in all hard and soft fractionated butterfats the ground pattern of the original butterfat is recognizable in the lower triglycerides C26-40 and the single SSM and double SMM unsaturated triglycerides of C44-54. The amount of the liquid oil which is high even in hard fractions of butterfat is due to it. The profile of the saturated (SSS) triglycerides with high carbon number C44-54 is on the contrary characteristically changed. The enrichment of the saturated triglycerides in hard fractionated butterfat was reason for trying to isolate them by an empirical crystallization method under such conditions that only the enrichment was determined.

Study of the separation of the saturated higher triglycerides in hard fractionated butterfat by crystallization. For the isolation of the SSS triglycerides of C44 and higher, the butterfats were crystallized in hexane at 12.5°C. Only in case of an enrichment of those higher melting triglycerides by fractionation was crystallization observed.

Our first experiments with repeated washings with cold hexane after a non-repeated crystallization procedure had a bad repeatability with variation coefficient of 10 to 30%. Introduction of the recrystallization and one washing by 4°C, gave an improvement with a V.C. of 5%. The method is an empirical one, fully defined by the described standardized method. No complete crystallization was obtained nor pursued.

Analysis of the composition of the obtained mass of crystals of the obtained filtrate and of the original fat material showed the expected fatty acid and triglyceride profile of higher saturated fatty acids and triglycerides (Table 5a and b).

| Table 5. Composition of the crystals | | | |
|---|---|---|---|
| a. Fatty acid profile | | b. Triglyceride composition | |
| C 4:0  0.1 | | C26  –  | C42  1.1 |
| C 6:0  0.1 | | C28  –  | C44  3.4 |
| C 8:0  0.1 | | C30  0.1 | C46  10.5 |
| C10:0  0.3 | | C32  0.1 | C48  19.1 |
| C12:0  1.0 | | C34  0.3 | C50  27.0 |
| C14:0  9.4 | | C36  0.6 | C52  19.6 |
| C16:0  40.1 | | C38  0.7 | C54  4.7 |
| C18:0  36.9 | | C40  0.7 | |
| C18:1  5.1 | | | |

The linearity of the method was tested by preparing a range of
mixtures with a fractioned and a non-fractionated butterfat.
At the chosen conditions, especially the temperature of 12.5°C,
the non-fractionated butterfat just gives no crystals. The method
is linear so this enables the analysis of butterfat samples
smaller than 25 grammes by filling up the sample with a non-fractionated butterfat.
The analysed hard fractionated butterfats gave 5 to 15% crystal
mass by the described method.
The detection limit of the method is 0.1% crystal mass.
75 Samples of non-fractionated (normal) butterfats from summer
and winter gave a crystal mass of mostly 0.0%. Only in two cases
0.2% was found.
Obviously, when more than 0.5% crystals is found, the occurrence
of fractionated butterfat is proved.

Interpretation of the percentage of the crystal mass. Fractionated
butterfat is a product that has not yet been defined. The average
amount of crystal mass we found was about 10%. When fractionated
butterfat was defined by this 10%, the detection limit of 0.5%
crystals means a detection limit of 5% hard fractionated butterfat of those average quality.
Where this method is only applicable for butterfats, the absence
of other animal or vegetable fats and in particular hydrogenated
fats must be certain. Otherwise an obtained mass of crystals may
be caused by these foreign fats in the butterfat. For the detection of these foreign fats in butterfat several international
standardized methods are available.

Detection of the soft fractionated butterfats. It will be clear
that, based on the same principle of the crystallization method,
soft fractionated butterfats can be detected when crystallization
is done at still lower temperatures. In that case the non-fractionated butterfat gives a big amount of crystals, whilst the
soft fraction gives none or a lower amount of crystals depending
upon the degree of fractionation. Where summer and winter butterfats differ considerably, the detection limit will be much higher than in case of hard butterfats.

Acknowledgement.

The authors wish to thank Mr Fokkema of the Butter Control Station of Friesland, Dr Ir Badings of the NIZO and Mr Smeets of The
Netherlands Voedselvoorzienings In- en Verkoop Bureau for their
willingness to make the butterfat materials available. Miss
Y. Louwers we thank  for her patience and care in preparing the
type-script.

Literature
[1] J.E. Schaap en G.A.M. Rutten. Voedingsmiddelentechnologie
(1974) 39, 8.
[2] J.E. Schaap en J.C. Kim. Voedingsmiddelentechnologie (1981) 14,
23.
[3] R.G. Black. Australian J. of Dairy Technol. (1973) 28, 116.
[4] B.G. Muuse, H.J. v.d. Kamp. RIKILT rapport 83.3 (1983).
[5] J.E. Schaap, H.T. Badings, D.G. Schmidt en E. Frede. Neth. Milk
Dairy J. 29 (1975) 242.
[6] H.T. Badings, J.E. Schaap, C. de Jong en H.G. Hagedoorn. Milchwissenschaft 38 (1983) 95.
[7] H.T. Badings, J.E. Schaap, C. de Jong en H.G. Hagedoorn. Milchwissenschaft 38 (1983) 150.
[8] J.M. de Man and M. Finoro. Can. Inst. Food Sci. Technol. J.
(1980) 13, 167.
[9] H. Timmen. 19e IDF congres B 7 (1974) 491.
[10] G. Lechat, P. Varchon, S. Kuzdzal-Savoie, D. Langlois,
W. Kuzdzal. Le Lait (1975) mai-juin no. 545-546 p. 295.
[11] Ch. Deroanne and A Guyot. Bull. Rech. Agron Gembloux (1974) 9,
261.
[12] P.W. Parodi. The Australian J. of Dairy Technol. March (1974)
20.
[13] B.G. Muuse, H.J. v.d. Kamp. RIKILT rapport 82.100 (1982).
[14] P.W. Parodi. J. of Dairy Res. (1981) 48, 131.
[15] P.W. Parodi. The Australian J. of Dairy Technol. March (1980)
17.
[16] H. Traitler, A. Prevot. J. of High. Resolution Chrom. and
Chrom. Comm. (1981) 4, 109.
[17] J.E. Schaap, G.A.M. Rutten. Neth. Milk Dairy J. (1974) 28, 166.
[18] H. Hendrickx, A. Huyghebaert, H. de Moor 19e IDF Congres (1974)
B 5, 230.

# Protein Analysis of Dairy Products P<sub></sub> by Ultrafiltration

By E. Renner

DAIRY SCIENCE SECTION, JUSTUS LIEBIG UNIVERSITY, BISMARCKSTRASS
GIESSEN, WEST GERMANY

## 1. Introduction

Of course the protein analysis of dairy products, which were
manufactured by ultrafiltration, is not completely different from
the protein analysis of other dairy products, as the same analytical methods are used. But, because the methods for the protein
determination in milk and milk products are based on the standard
procedure according to Kjeldahl for the determination of nitrogen
it has to be respected that the ratio of protein nitrogen to non-
protein nitrogen (NPN) can be changed to a very high extent during
the ultrafiltration of milk or whey. Therefore, it is necessary
that a procedure for the protein determination is used which permits a differentiation of these components. Furthermore, these
dairy products often contain an increased proportion of whey protein, which influences the biological protein value [1]. Therefore,
it seems also to be necessary as to the protein analysis to differentiate between the proportions of casein and whey protein.

## 2. Analytical procedure

I would like to describe the analytical procedure at the example
of ultrafiltration quarg, as it is manufactured in an increasing
extent in the Federal Republic of Germany. By using such a procedure also the whey protein is passing in a high percentage into
the fresh cheese; furthermore it has to be respected that because
of the heat treatment of the cheese milk a part of the whey protein is bound to the casein and will therefore also pass into the
quarg.

In Figure 1 it is shown  how a separation of the individual protein and nitrogen fractions in the quarg can be achieved. 30 g of
quarg are centrifuged. In the quarg concentrate, the complex

_ormation between casein and whey protein is solved by adding an emulsifying salt solution. Then the casein is precipitated again by acetic acid, so that after the centrifugation the nitrogen in the concentrate II represents the casein content, whilst the nitrogen in the solution represents the whey protein which was originally bound to the casein.

If in the whey which occurs after the centrifugation of the quarg the protein is precipitated by trichloroacetic acid the nitrogen of the precipitate can be characterized as the originally soluble whey protein of the quarg, whilst the solution contains the NPN. By adding both whey protein values the proportion of total whey protein in the quarg can be achieved.

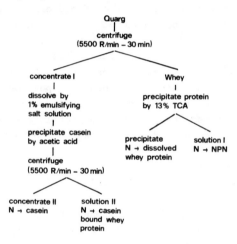

Figure 1: Analytical procedure for the determination of the pro-
        tein resp. nitrogen fractions of UF quarg

In Table 1 the results of such an analytical procedure are presented for the example of an UF quarg sample. The quarg sample contains 14.2 % protein with a proportion of 85.1 % casein and 11.9 % whey protein, which is the sum of the dissolved and of the bound whey protein of 6.2 resp. 5.7 %. The whey contains 30.1 % of the nitrogen as NPN.

Table 1: Protein composition of an UF quarg sample
          (WP = whey protein)

```
Total N:   2.218 %        ──→ 14.2 % protein
Casein N: 1.888 %         ──→ 85.1 % casein of total N
Whey: WP:  0.138 %        ──→  6.2 % WP of total N (I)
      NPN: 0.059 %        ──→ 30.1 % NPN of N in whey
Solution II: WP: 0.126 % ──→  5.7 % WP of total N (II)
Whey protein (I + II) = 11.9 % of total N
```

## 3. Results

By using a few examples, it will be demonstrated how on the
basis of such an analytical procedure the changes can be re-
gistered in dairy products which were manufactured by ultrafiltration.

In Table 2 the protein composition of quarg samples is compared
which were produced by different procedures: according to the
traditional procedure, according to a so-called thermo procedure,
where because of a heat treatment an intensified complex formation between casein and whey protein and therefore a higher yield
is achieved, and according to the ultrafiltration procedure.
There is a somewhat higher protein content in the UF quarg but
not in a significant extent. The changes of the protein composition, in the first line, are characterized by an increased proportion of whey protein of total nitrogen and by a reduced proportion of casein in the thermo as well as in the UF quarg. From
these results a higher biological protein value of these quarg
samples can be derived.

Which changes in the nitrogen composition can occur during the
ultrafiltration procedure  can be seen in Figure 2. During the
ultrafiltration of skim milk the NPN proportion of total N is decreasing from 6 to about 1.5 %, during the ultrafiltration of
whey from 32 to 7 %. That means that the ratio true protein/NPN
in skim milk is increased from 16:1 to 65:1 and in whey from
2.1:1 to 13:1.

In particular, such changes are very important  when a balance
is to be made in order to register exactly the utilization of the
milk protein for the individual milk products. If for the ultra-

Table 2: Influence of the manufacturing procedure on the protein
content and protein composition of quarg [2]

| Procedure | Protein content | Proportion of | | | | Bio-logical value |
| | | casein | whey protein | bound whey protein | NPN in whey | |
| | % | % | % | % | % | |
|---|---|---|---|---|---|---|
| Traditional | 12.7 | 92.4 | 6.1 | 2.2 | 27.2 | 81 |
| Thermo | 13.0 | 89.6 | 8.3 | 4.0 | 30.8 | 83 |
| Ultrafiltration | 13.8 | 87.5 | 9.5 | 4.0 | 33.3 | 84 |
| F value *) | 2.31 | 13.0 | 10.1 | 8.80 | 2.98 | 10.0 |
| | - | *** | *** | *** | - | *** |

*) Statistical significance: - = not significant
                              *** = highly significant

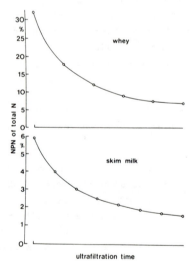

Figure 2: Influence of ultrafiltration on the NPN content of skim
milk and whey

filtration of skim milk and whey such a balance is performed on
the basis of total nitrogen, it will become incorrect, as thereby
a transfer rate of 5 % protein from skim milk and of about 30 %
from whey into the permeate is expressed (Table 3), which is not
true, because it is not protein but exclusively NPN. In fact, all
the true protein remains in the retentate.

Table 3: Passage of total nitrogen and true protein from skim
milk and whey into the retentate and permeate by ultrafiltration [3]

| Product | Distribution product | Transfer rate (%) of | |
|---|---|---|---|
| | | total nitrogen | true protein |
| Skim milk | retentate | 94 | 98 |
| | permeate | 5 | - |
| Whey | retentate | 71 | 96 |
| | permeate | 29 | - |

I think that these examples make it clear, that a correct protein
analysis of dairy products which were manufactured by ultrafiltration is very important.

References
[1] E. Renner, "Milk and Dairy Products in Human Nutrition", Volks-
wirtsch. Verlag, Munich, 1983.

[2] E. Renner, U. Karasch, A. Renz-Schauen and A. Hauber, Deut.
Milchwirtsch., 1983, 34, 1410.

[3] G. Rommel, "Zur Eiweissbilanzierung in milchwirtschaftlichen
Betrieben", Thesis Justus Liebig University, Giessen, 1983.

# Residues and Contaminants in Milk and Milk Products

By A. Blüthgen, W. Heeschen*, and H. Nijhuis

INSTITUTE FOR HYGIENE, FEDERAL DAIRY RESEARCH CENTRE, P.O. BOX 1649, D-2300 KIEL, WEST GERMANY

## Introduction

As a result of the increasing standards concerning both the quality and the quantity of foods of animal origin and in particular of milk a great number of chemical compounds is used either directly or indirectly in the production. In addition, the position of this branch of industry leads, in a more or less contaminated environment, to undesirable conflicts of goals between producers and consumers which are frequently difficult to overcome and only by compromises.

Today the consumer has become particulary sensitive as regards adverse changes in his food. This may mainly be attributed to 3 concurrently observed trends:

- The economic need for rationalization in the agricultural field has led to the increased use of pesticides and active compounds for plants and animals compared with former times.

- As a result of expanding industry and increasing traffic naturally occurring heavy metals as lead, cadmium and mercury, for which physiological functions are not known and which are therefore regarded as toxic heavy metals, are set free to an extent exceeding by far geological mobilization and are emitted into the environment and also the food chain. Toxic heavy metals can also secondarily contaminate the milk after its production.

- The rapid development of modern microanalytical techniques permits components and impurities in foods to be increasingly detected by using methods which become more and more sensitive. Today the limit of detection of trace contaminants is within the range of micrograms and nanograms (ppb and ppt, resp.).

Contamination of milk with undesirable residues may occur through the animal itself, the environment in a broader sense and also during further processing. For defined areas the residues to be expected are more or less characteristic,

although their concentration may vary considerably. The extremely complicated
relationships are presented in Table 1[1].

Table 1:  Milk and milk products with residues of chemicals of different origin

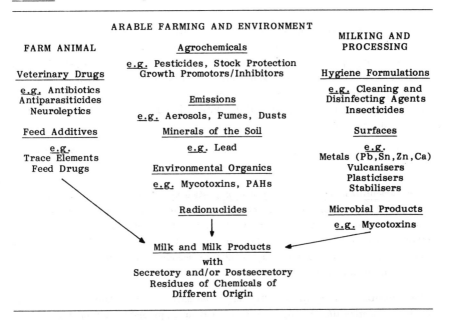

From Table 1 it may be seen that contamination of milk may occur either **via**
the animal (secretory) or following milking (post-secretory) through residues of
chemical substances of largely varying origin and categories, the process of contamination being mainly influenced by the biological phenomenon of the food
chain. So certain contaminants may enrich from step to step, such as the persistent chlorinated hydrocarbon pesticides, whereas the toxic heavy metals as
lead and cadmium are stored mainly in the skeleton or in kidney and liver.

Basic terms[2,3]

It is essential to use only clear, well established definitions to describe problems, associated with residues of drugs, biocides (pesticides) and other foreign
substances. The same applies for various aspects of the evaluation of the toxicological significance of the residues. Therefore, some terms used in residue analysis and interpretation  have to be explained.

A pesticide is any substance or mixture of substances intended for preventing or controlling any unwanted species of plants and animals and also includes any substance or mixture of substances intended for use as a plant-growth regulator, defoliant or desiccant (e.g. fungicides, insecticides, parasiticides, herbicides, molluscicides).

A foreign substance or contaminant can be defined as any substance, not intentionally added to food, which is present in such food as a result of production (including operations carried out in crop husbandry, animal husbandry and veterinary medicine) or in manufacture, processing, preparation, treatment, packing, packaging, transport or holding of such food or as a result of environment contamination (Aflatoxin $M_1$, PCBs, metals, cleaning and disinfecting agents, drugs).

A pesticide residue is any substance or mixture of substances in food for man or animals resulting from the use of a pesticide.

Good agricultural practice in the use of pesticides (GAP) is the officially recommended or authorized usage of pesticides under practical conditions at any stage of production, storage, transport, distribution and processing of food and other agricultural commodities, bearing in mind the variations in requirements within and between regions and taking into account the minimum quantities necessary to achieve adequate control.

The acceptable daily intake (ADI) of a chemical is the daily intake which, during an entire lifetime, appears to be without appreciable risk on the basis of all known facts at the time. It is expressed in milligrams of the chemical per kilogram of body weight.

A maximum residue limit (MRL) is the maximum concentration of a pesticide residue in mg/kg (ppm), resulting from the use of a pesticide according to good agricultural practice directly or indirectly for the production and/or protection of the commodity for which the limit is recommended.

An extraneous residue limit (ERL) is, for a particular commodity, the maximum toxicologically acceptable concentration of a residue unavoidably arising from sources other than the use of a pesticide directly or indirectly for the production of that commodity.

A guideline level is the maximum concentration of a pesticide residue that might occur after the officially recommended or authorized use of a pesticide for which no acceptable daily intake or temporary acceptable daily intake is established and that need not be exceeded if good practices are followed.

Basis for evaluation of chemical residues and contaminants

The effects injurious to health of toxic foreign substances are normally determined on the basis of the results obtained from extensive trials with animals
using varying species.

By elucidating the relationship between dose and effect that amount of compound may be defined which allows no longer a measurable effect to be detected
using sensitive methods (control of growth of young organisms, clinical, histo-
logical, hematological and enzymological examinations). This limit value is also
termed non-effect level (NEL) and is expressed in mg/kg of body weight.

To establish the acceptable daily intake (ADI) of a toxic foreign substance, the
results obtained from the trials with animals are extrapolated taking a safety
factor of normally 100 into account, and up to 1,000 where particular reason for
concern exists. Therefore, ADI is defined as the proportion of a chemical residue, expressed in milligram per kilogram of body weight, which, for a long
period of time, may be taken in by the consumer without appreciable risk on
the basis of all of the known facts at the time. In this context, the term
"without appreciable risk" means that the indicated daily dose does not lead to
injuries to health even when taken during an entire lifetime.

Max. levels (maximum residue limits) are a sort of warning, either reaching or
exceeding them being indicative of, for instance, disorderly application of a
chemical substance. Naturally, there remains still a risk which cannot be calculated or is difficult to calculate even when the afore-mentioned risks are observed, since there is no absolute safety in biology. This remaining risk is,
among other things, attributable to the fact that there exists no true relationship between dose and effect as regards numerous serious effects. No activity
range can, for instance, be stated till now for carcinogenic substances. This
concerns also nitrosamines or defined mycotoxins. In these cases it is not possible to establish maximum levels on an exact scientific basis. Taking into account the acceptable daily level, the body weight and the amount of foodstuff
consumed it is possible to establish "permissible concentrations"(PL). This is
done according to the following formula:

$$(PL) = \frac{ADI \ (mg/kg \ body \ weight) \ x \ body \ weight \ (kg)}{amount \ of \ foodstuff \ (kg)}$$

Veterinary drugs

Problems associated with residues of drugs in milk may occur particularly where
defined drugs must be applied on a large scale (treatment of whole stocks).

Application of drugs for dairy cows, however, is subject to so-called waiting times or withholding periods which are officially established. Observance of these waiting times can normally be expected of the milk producer even from the economic point of view. For drugs there exist frequently general waiting times, e.g. 5 days, or special waiting times established for a single drug and which may vary between 0 and several weeks. With defined residues of drugs (e.g. antibiotics) one has to bear in mind the dairy's self-interest, because the so-called inhibitors may cause considerable problems in the production of milk products made by using starters and starter cultures. Therefore, the dairies frequently carry out analyses for inhibitors on their own responsability using extremely sensitive methods.

Chemotherapeutics (antibiotics and sulfonamides)[4]. The most important antibiotics, applied to the lactating cow, are penicillin, streptomycin, tetracyclines, chloramphenicol, neomycines and some others. Sulphonamides are often used in combination with antibiotics.

Indications for the use of antibiotics and sulphonamides in dairy cattle include therapy and prophylaxis of diseases of primary and secondary nature caused by microorganisms. According to locality and intensity of the disease different forms of application are used, and a great number of veterinary-pharmaceutical preparations are available. The intramammary application of these drugs in mastitis therapy is the most important source of milk contamination.

The concentration of antibiotics in milk after intramammary application and the duration of excretion are influenced by the following factors:

- dosage
- time interval between treatment and first milking
- absorbency of the udder tissue (pathological-anatomical changes)
- milk yield and individual factors

Sulphonamides are mostly used in the treatment of acute bacterial mastitis and are applied primarily intravenously because of their tissue-irritating effect. Only a relative small number of sulphonamides is suitable for intramammary application.

For residue analysis physico-chemical procedures and microbiological methods are available.

Table 2 gives a summary of the physico-chemical procedures used for the final determination of antibiotic residues in milk without considering the preparation of the samples prior to measurement[4].

Table 2 :  Physico-chemical procedures for the final determination of antibiotic
residues in milk

| Procedure | Determination of | Sensitivity | Principle of application |
|---|---|---|---|
| Photometry | Penicillin | 10.0  mg/kg | Measurement, limited identification |
| Electrophoresis including microbial detection | Penicillin<br>Tetracycline | 0.02  IU/ml<br>5.0  IU/ml | Limited measurement identification |
| Gas chromatography | Isoxazolylpenicillin | 0.01  IU/ml | Measurement, limited identification |
| Thin layer chromatography, Detection in UV-light | Tetracycline | 0.025 mg/kg | Limited measurement, identification |
| Polarography | Chloramphenicol | 3.0  mg/kg | Measurement, limited identification |
| Thin layer chromatography | Chloramphenicol | 0.5  mg/kg | Limited measurement, identification |
| Gas chromatography | Chloramphenicol | 0.001 mg/kg | Measurement, identification |

Chromatographic methods are suitable for the separation of mixtures of antibiotics and thus allow the individual substances to be identified. Apart from thin layer chromatography and column chromatography, gas chromatography has recently been used as a quantitative method of determination in the sub-microgram range.

Fluorometry does not permit the separation of a mixture during the final determination and is therefore limited to the analytical identification of isolated antibiotics. The sensitivity is relatively high.

Polarographic and photometric procedures are also used for the determination of antibiotics in milk.

The principle of detecting inhibitory substances in milk by microbiological methods consists in assessing any changes in growth and/or metabolic activity of sensitive microorganisms due to the presence of inhibitors as compared with the control milk definitely free of inhibitors. Table 3 gives a summary of the microbiological methods for the detection of inhibitors in milk[4]:

Table 3: Microbiological methods for the detection of inhibiting substances in milk

| Principle | Method | Test organism | Detection limits/ml | | Incubation time at opt.temp.(h) |
|---|---|---|---|---|---|
| Agar diffusion | Disc assay | Bac.subtilis (25-37°C) | 0.005-0.02 IU Penicillin | | 4 - 6 |
| | Disc assay | Bac.stearo-thermophilus (55-60°C) | 0.0025 IU Penicillin | | 2 1/2 |
| | Disc assay | Sarcina lutea A.T.C.C. 9341 (28°C) | 0.005 IU Penicillin | | 6 - 8 |
| | Brilliantblack reduction test | Bac.stearo-thermophilus var.calidolactis C953 (55-60°C) | 0.006 IU Penicillin 0.8 y Chlortetra-cycline 8 y Streptomycin | | 2 1/2 |
| Dye reduction | TTC test | Str.thermophilus (37°C) | 0.03 IU Penicillin 0.2 y Chlortetra-cycline 5 y Streptomycin | | 2 1/2 - 3 |
| | Methylenblue | Str.thermophilus (+ L.bulgaricus) (37°C) | 0.02 IU Penicillin 3 y Streptomycin | | 2 1/2 - 3 |
| Inhib. of acid formation | Litmus | Str.thermophilus (+L.bulgaricus) (40°C) | 0.03 IU Penicillin 0.5 y Tetracycline 10-50 y Streptomycin | | 2 1/2 - 3 |
| | Bromcresol purple | Str.thermophilus (45°C) | 0.01- 0.05 IU Penicillin | | 6 - 22 |
| | Coagulation test | Str.thermophilus (40°C) | 0.02 IU Penicillin 0.5 y Chlortetra-cycline 6.5 y Dihydro-streptomycin | | 16 |
| | Titration method | Str.thermophilus (+ L.bulgaricus) (40°C) | 0.008-0.01 IU Penicillin 0.2 y Chlortetra-cycline 0.6 y Streptomycin | | 5 |
| Changes in the growth density | Biophoto-metric method | Staph. aureus Oxford (37°C) | 0.005 IU Penicillin 0.001-0.05 y Chlor-tetracycline 0.1 y Streptomycin | | 1 1/2 - 3 |
| Morpho-logical changes | Microscopic detect. of morph.changes | Str.thermophilus (37°C) | 0.015 IU Penicillin | | 1 |

Note: 'y' should read µg/g or mg/kg

Microbiological methods show in many cases high sensitivity, involve small expenditure of personnel and apparatus and simple operation. However, it is not possible, with microbiological methods, to reveal the presence of a particular antibiotic, except in the case of the penase test which allows penicillin to be determined in milk. Using different test microorganisms it can be possible to have a preliminary identification of a certain antibiotic.

Due to the possible occurrence of native inhibitors in milk of single cows most microbiological inhibitory tests should be applied only for bulk milk or in cases, where the excretion time of a certain antibiotic should be determined.

The number of inhibitor-positive raw milk samples varies from country to country and depends on the fact, whether in the respective country inhibitory tests are regularly carried out for quality grading or food inspection purposes. In countries where inhibitory tests are done regularly, the frequency of inhibitor-positive raw milk samples in herd bulk milk is not higher than 2-4/1,000. Other reasons for different numbers of inhibitor-positive milk samples may be varying sensitivity of the methods used and different patterns of antibiotics, which vary from country to country. Also seasonal influences on the number of inhibitor-positive samples are known. In general it can be said that a drastic reduction of inhibitor-positive results is observed in countries after introduction of routine tests for inhibitors[4].

The significance of antibiotic residues in milk has to be regarded with respect to the economic losses, due to interference with manufacture of dairy products and with respect to food hygienic aspects.

Milk containing antibiotics gives rise to problems in the acidification and ripening of cheese, acidification and flavour development in butter, buttermilk, yoghourt, and cultured milk products. In addition, quality controls such as a reductase test and bacteriological examinations may indicate considerably better quality than actually exists.

Heating processes like pasteurization have little or no influence on the antibiotics present in milk. Chloramphenicol is completely heat resistant. The sensitivity to temperature increases in the order: chloramphenicol, penicillin, streptomycin, tetracyclines. Boiling or heating milk to 100°C destroys antibiotics to the extent of 50 % (penicillin), 66 % (streptomycin), and 90 % (chlortetracycline and oxitetracycline)[4].

It is considered that possible harm by antibiotics in milk may be produced in three ways:

- direct toxic effects of antibiotic residues
- induction of allergies
- development of resistance

Direct toxic effects from the ingestion of milk containing antibiotics are not clearly stated. With the very small concentrations of antibiotics in milk no sensitising effect can be expected and none has been described so far[5].
Resistant organisms are not expected to occur after occasional and low intakes of antibiotics.

Fasciolicides[6]. In many countries, fasciolosis caused by the common liver fluke, Fasciola hepatica, is, from the economic angle, one of the most important parasitic diseases of cattle. Systematic measures for controlling fasciolosis in cattle are therefore of direct and high interest to the owner of the animals. Generally treatment of the whole herd is recommended including the dairy cows. Fasciolicides are normally applied in amounts of 3-10 mg/kg of body weight.

Treatment of the host (cattle) with fasciolicides should be supported by the eradication of the intermediate host, which in Europe is Galba trunculata. The most effective molluscicide is sodium pentachlorophenolate, but since treatment must be applied to and around waterways, the provisions of water supply laws have to be kept in mind[7].

The most important fasciolicides, their application, dosage and effectiveness in cattle are given in Table 4[6].

Table 4: Fasciolicides - application, dosage, and effectiveness in cattle

| Feature | | | Effectiveness | |
|---|---|---|---|---|
| Active substance | Application | Dosage mg/kg body weight | immature stages | mature stages |
| Bromphenophos | oral | 12 | good effect with therapeut. dosage | 85% - 100% |
| Niclofolane | oral | 3-5 | partial effect with high dosage | 85% - 97% |
| Nitroxynil | subcutaneous | 10 | good effect with 1,3-fold dosage | 93% - 100% |
| Oxyclozanide | oral | 10 | moderate effect with 4-fold dosage | 83% - 88% |

Information is limited about <u>metabolism and excretion</u> of the various fasciolicides. Essentially they are excreted with faeces and, in a combined form, with the urine and in lactating cows also with the milk. Depending on the type of drug, application of a fasciolicide to lactating cow may produce changes in the biochemical status of the milk as well as residues.

Table 5 summarizes the excretion of fasciolicides in milk after the therapeutic doses. The data were developed from trials with animals giving comparable milk yields of the order of 10-12 kg/day[6,8].

<u>Table 5:</u> Excretion of fasciolicides in milk of cows after application of therapeutic doses

|  | Niclofolane | Oxyclozanide | Nitroxynil | Bromphenophos |
|---|---|---|---|---|
| Peak concentration excreted (mg/kg) | 0.06 | 0.13 | 1.46 | 1.5 |
| Time (hrs) until the analytical detection limit is reached (0.01 mg/kg) | 168 | 168 | 216 | 144 |
| Ratio of dose to concentration in milk at end of 120 hours (partly extrapolated) | 600:1 | 500:1 | 56:1 | 400:1 |
| Ratio of concentration in milk at end of 120 hours to 0.01 mg/kg | approx. 1:1 | approx. 2:1 | approx. 20:1 | approx. 1:1 |
| Influence on milk quality for 120 hours | Coagulation and hardening time increased | No influence on suitability for processing in the dairy | None | pH-value and chloride content slightly increased |

The period during which treated cows excrete residues is clearly higher with Nitroxynil than with the other compounds.

<u>Residues of fasciolicides</u> in milk can be determined by different analytical methods. The method of choice is gas-liquid-chromatography. Additional identification can be performed by mass spectrometry or by photometry.

The extraction of residues from milk is performed by a multistep liquid-liquid-partioning between the aqueous medium of certain pH and a suitable organic solvent.

This leads in most cases to a concentrate, suitable for GLC. The phenolic nature of the active compounds makes a derivatization step necessary.

Niclofolane, Nitroxynil and Oxychlozanide are methylated with diazomethane in ether, while Bromphenophos is silylated with BSA.

In all cases the ethers which are formed may be detected and quantified in the mg/kg range on gaschromatographic columns of moderate to low polarity. Due to the high halogen-content, the electron-capture detector, preferably fitted with a nickel 63 foil, which gives a tentative identification is recommended[9].

Ectoparasiticides. Control of ectoparasiticides in the lactating animal is indispensable particularly during the winter months in order to prevent considerable disturbance of the animals by lice, flies and others. Application of chlorinated hydrocarbons such as lindane ($\gamma$-HCH), even in small quantities, is likely to cause violations of tolerance limits[10].     Despite their good effect lindane (or technical HCH) should no longer be employed in parasite control in lactating cows due to the considerable risk of residue formation. Also when lindane is applied to non-lactating animals one has to bear in mind that considerable depots of residues may build up in the animal which are excreted after beginning of lactation.

For ectoparasite control in the lactating animal phosphoric esters (Coumaphos, Trichlorfon and Heptenophos) can be recommended. In some cases waiting times need not be observed (waiting time = 0). Further, Permethrin compounds may be used. It is not clearly known whether application of Permethrin-containing compounds may lead to manifestation of resistance of ectoparasites.

Organophosphate-insecticides are not only used widely against the various ectoparasites, but due to their systemic action also against fly larvae, which affect not only the health of cattle, but also the economy of the dairy industry.

Ectoparasiticides may leave residues in milk, the concentrations depending on the insecticide and the method and rate of application. In Table 6 a variety of the most important compounds used, the acceptable daily intakes, the permissible levels and the residues after application of therapeutic doses are listed.

Table 6: Ectoparasiticides and their residues in milk

| Compound | ADI mg/kg BW | Permissible level (1 kg milk) | Residues (mg/kg milk) Max. | 120 h |
|---|---|---|---|---|
| Bromocyclen | 0.05 | 3.0 | 0.008 | 0.001 |
| Coumaphos | 0.0005 | 0.03 | 0.01 | 0.005 |
| Trichlorphon | 0.01 | 0.60 | 0.02 | 0.005 |
| Phoxim | 0.0002 | 0.012 | 0.05 | 0.01 |
| Lindane | 0.01 | 0.60 | 0.40 | 0.012 |
| Permethrin | 0.05 | 3.0 | 0.03 | 0.012 |
| Chlorpyriphos | 0.001 | 0.06 | 0.28 | 0.02 |
| Bromophos | 0.003 | 0.18 | 0.10 | 0.011 |
| Fenchlorphos | 0.01 | 0.60 | 1.20 | 0.013 |

For residue analysis the chlorinated hydrocarbons such as lindane can be extracted from the milk with methods which are also used for other chlorinated hydrocarbons. Determination is done by gaschromatography, using an electron-capture detector.

Residues of organophosphates and carbamates in milk are determined most conveniently by gas-liquid chromatography (GLC). Although the thin-layer chromatography has been considerably improved as far as limits of detection, instrumentation and general technique are concerned, the GLC-method is the method of choice. With carbamates the GLC-determination of residues is not always easy. Recent progress in instrumentation (specific detectors) as well as in derivatization of carbamates, however, opened new ways in residue analysis of these compounds but even so it is difficult to determine the water-soluble metabolites of carbaryl and other carbamates in milk. It has been accepted that measurement of the parent, carbaryl, in milk provides a satisfactory means of judging whether good agricultural practices have been followed and thus gauging the public health implications of any residues that may be present[11].

GLC analytical methods for residues usually comprise three steps; extraction of the residues from the matrix, clean-up of the extract by partition and/or column chromatography with subsequent elution of the residues and detection of the residues in eluates by means of a suitable detection system.

In case of milk and milk products, acetonitrile partition with subsequent Florisil clean-up, followed by elution of the organophosphates and their metabolites by solvents of increasing polarity, constitute the major steps of the analytical pro-

cedure for residues. For detailed description of the analytical methods for individual organophosphates and carbamates, specific literature or journals should
be consulted.


<u>Teat disinfectants (iodophors)</u>[12,13]. During the last decade considerable interest
has arisen in the significance of post-milking teat dipping as a hygienic measure
for teat skin protection and the prevention of new infections of the bovine
mammary gland[14].

From the angle of teat skin protection the need for teat dipping arises from the
changed conditions of milk production. The milking ointments used during
mechanical milking may have an adverse effect on the liners. Teat disinfection
in mastitis herds is required, for even under favorable hygienic conditions the
spread of pathogens cannot be prevented during milking using pulsator, pulsat-
ing liner and claw.

For teat dipping a variety of active ingredients and combinations of active ingredients containing a disinfecting component and an emollient are recommended,
either to kill the pathogens left on the teats and the teat orifices or to prevent
their growth after each milking. In many countries particular attention has been
given to the use of the so-called <u>iodophors</u>, which were developed in 1950. The
germicidal activity of the iodophors is due to the dissociation of the iodine of
the surface active agent at dilution. Efficacy does not vary markedly at temperatures ranging between 20-37°C. Iodophor teat dips are used in concentrations of 1,000-10,000 mg/kg, the normal concentration being about 5,000
mg/kg. Active chlorine preparations contain 1-4 % active chlorine. Despite their
low cost, they are accepted to a varying extent only. Quite often, they have
been blamed for teat skin irritation. Further, it has not been possible until now
to prepare a stable solution with an appropriate pH neither for immediate
application nor for use following dilution. Other active ingredients with skin
disinfecting properties are hexachlorophene, quaternary ammonium compounds
and chlorhexidine; however, they have not gained practical importance. Glyce-
rine (up to 10 %) and lanolin (up to 3 %), for instance, are used as emollients.
After application of teat dips iodine residues in milk might be of importance[15].


The <u>physiological iodine levels in milk</u> vary between nearly 0 and 700 µg/kg
with averages within the range of 20-200 µg/kg and mean values between 40 and
80 µg/kg. The main influence on the iodine content of milk is due to the iodine
content of feeding stuffs[16].

The regular use of iodine-based teat dips containing concentrations of 1,000 to 10,000 mg/kg available iodine is associated with a marked increase in the iodine levels in milk (Table 7)[10].

Table 7:          Increase of iodine content in milk after use of iodophors

| Iodine level of dip solution (mg/kg) | Increase of iodine content in milk (µg/kg) |
|---|---|
| 1.000 | 23 |
| 2.500 | 19 |
| 3.300 | 53 |
| 3.750 | 77 |
| 4.500 | 85 - 130 |
| 5.000 | 33 - 420 |

The increase in iodine in herd bulk milk should be normally after regular teat dipping not more than 150 µg/kg. This level can be kept with dip preparations, containing not more than 3,000 mg available iodine/kg.

For the detection of iodine in milk several principles of detection with varying degrees of specificity and sensitivity are available. Using an analytical procedure one has to distinguish between determination of total iodine, protein-bound iodine and iodides in milk. Determination of the latter substances are often considered to be sufficient, as the iodides form the greater part of the total iodine (90 %).

Trace levels of iodide have been determined by the following methods

- Neutron activation analysis
- Catalysis of the reaction between ceric salts and arsenious oxide
- Other catalytic methods
- GLC techniques
- Atomic absorption spectrophotometry
- X-ray fluorescence
- Specific ion electrodes

The following Table 8 compares the attributes of various methods for determination of iodide in milk.

Table 8:        Determination of iodide in milk[10,15,16,17]

| Procedure | Limit of Detection (µg/kg) | Recovery (%) | Precision (%) |
|---|---|---|---|
| Ce IV - AS III | 10 | 94 - 103 | - |
| AAS | 0.2 | - | - |
| X-ray fluorescence | 100 | - | $\pm$ 10 |
| GLC | 10 | 80 - 100 | 3 - 10 |
| Titration | (mg/kg) | 93 | 0.6-3.3 |
| Specific ion electrode | 40 | 101 $\pm$ 7 | 3.4 |

Iodophor teat dips normally contain detergents (tensides), especially alcylphenol-ethylene-oxid-condensation products (nonoxinols). 15-30 g of these tensides are included in teat dips. In most of the teat dips nonoxinol 9 or 15 are used.

Thin layer chromatographic investigations for nonoxinol 9 in milk have shown that there is normally less than 1 mg/kg milk present. These findings are in agreement with those of SCHUMACHER[15], who could not find nonoxinol 15 residues of more than 1 mg/kg. In Switzerland for nonoxinol 15 a tolerance level of 2 mg/kg milk is in operation.

For toxicological evaluation of iodine residues in milk it is important to realize that according to the recommendations of WHO, the optimum daily intake of iodine should be 150 to 200 µg/person. In a number of countries, the daily intake actually does not exceed 30 to 70 µg/person. As a consequence of a dietary iodine lack thyromegaly may occur endemically.

In the light of the present knowledge of iodine residues in milk, a daily intake exceeding 500 to 700 µg/person is undesirable and may cause hyperthyroism. Under normal conditions, that means the daily consumption of max. 1 l/person, an excessive uptake of iodine appears scarcely to be possible, even if teat dipping is practised extensively.

The toxicological relevance of the tensides was extensively studied by SMYTH et al.[18] The non-effect-level for nonoxinol 9-15 was calculated to be in the range between 30 and 100 mg/kg. Carcinogenic effects could not be proved in any case. With nonoxinol 20 in dogs cardiotoxic effects could be observed. These effects could not be stated with nonoxinols 12 or 25 respectively.

## Chlorinated hydrocarbons ("Pesticides")[19]

Since the discovery of the insecticidal properties of DDT in 1940, a number of organochlorine insecticides have been developed. These are characterised by differing persistency, toxicological properties, specific effectiveness and residual behaviour.

The use of organochlorine pesticides has been forbidden or restricted in the neighbourhood of milking animals in numerous countries[20].

Intentional or unintentional contamination of the lactating cow may arise from the following sources:

- Control of parasites on the animal
- insect control in stables
- feeding stuffs
- contamination from the environment (water, air, soil)
- contamination by udder hygiene preparations containing wool wax (lanolin)
- accidents, negligence and others.

Feeding stuffs might be the most important sources of contamination especially in areas where high amounts of concentrates are fed. The possible intake of organochlorine residues might be due to oilseed cake (cottonseed, linseed, sunflower, peanut), grain (maize, barley, oats, sorghum, wheat), milling offals (bran, shorts, silo dust), food wastes (citrus pulp, apple pomace, sugar beet-pulp, grape pomace, potato peel), root crops (sugar beets, feed beets, potatoes), meat meal and animal fat.

Both the cutaneous contact with organochlorine pesticides and the intake of pesticides contained in fodder and soil give rise to residues in the body-fat of the animal and in the milk. The concentration of such residues depends not only on the amount ingested or applied, but also on the pesticide compound involved.

Because of their lipophilic nature and their relative stability, most of the chlorinated insecticides and their metabolites are excreted into the milk fat. Lactating female animals rid themselves of residues significantly more rapidly than males of similar weight, age and residue content. The water-soluble metabolites are excreted *via* urine and faeces. The rate of excretion into the milk depends on several factors including stage of lactation of the animal, the quantity of milk fat produced daily and the breed of the cow, the nature of the pesticide, the amount consumed daily, the duration of exposure, previous history of exposure to organochlorine pesticides and others.

An important problem in many countries is the occurence of ß-HCH in the milk after ingestion of small amounts of ß-HCH with the feeding stuff. Recent investigations have shown that ß-HCH can have excretion percentages of up to 80 and 100. If a cow is fed with concentrates containing 0.01 mg/kg feeding stuff (dry matter), the total intake will be 0.1 mg ß-HCH. If this cow produces daily about 1 kg of milk fat and the carry over reaches 100 %, it cannot be excluded that, together with other contamination sources, a contamination of the milk fat of 0.1 mg/kg is obtained. In many countries this figure is near the maximum residue level, fixed by tolerance-regulations[21]. In certain areas with very intensive milk production amounts of 10-15 kg concentrates are fed daily, and this might be in many cases the reason for elevated concentrations of ß-HCH.

Since many years it is well-known that α- and ß-HCH possess no insecticidal properties. This is only the case with γ-HCH (lindane). During production of HCH and lindane the so-called technical HCH is formed, which presents a mixture of the different isomers. For economic reasons technical HCH is still used in some cases for plant protection, environmental hygiene, protection of crops, control of parasites on animals. It is a very severe task for the future to prevent the further application of technical HCH.

The following Table 9 gives an idea of average organochlorine residues, found in milk in different countries.

Table 9:   Organochlorine residues in milk (mg/kg on fat basis)[2]

| Pesticide | Min-max values | Average residue level |
|---|---|---|
| HCB | 0.01 - 0.5 | 0.05 |
| γ-HCH | 0.00 - 0.3 | 0.02 |
| α-HCH | 0.00 - 0.1 | 0.02 |
| ß-HCH | 0.00 - 0.1 | 0.03 |
| Dieldrin | 0.00 - 0.03 | 0.005 |
| Heptachlorepoxid | 0.00 - 0.03 | 0.005 |
| DDT | 0.00 - 1.00 | 0.05 |

For the <u>determination of residues of organochlorine pesticides</u> in milk and milk products an IDF-Standard (75A:1980)[22] is available, which is now under revision. The methods described are guidelines for use in actual laboratory practice. They can be considered as equivalent.

The methods A-H can be described according to the clean-up-principles as follows:

Method A:  Liquid-liquid partitioning with acetonitrile and clean-up on a Florisil column

Method B:  Liquid-liquid partitioning with dimethylformamide (DMF) and clean-up on an alumina column

Method C:  Liquid-liquid partitioning with dimethylformamide (DMF) and clean-up on a Florisil column

Method D:  Column chromatography on aluminium oxide

Method E:  Column chromatography on alumina column

Method F:  Column chromatography on partially deactivated Florisil

Method G:  Column chromatography on partially deactivated Silica gel

Method H:  Gel permeation chromatography

In Table 10 it is shown, for what compounds the different methods are applicable.

Table 10: Applicable methods for the determination of organochlorine pesticides

| Compound | Method | | | | | | | |
|---|---|---|---|---|---|---|---|---|
| | A | B | C | D | E | F | G | H |
| α-HCH | + | + | + | + | + | + | + | + |
| ß-HCH | + | + | + | + | + | + | + | + |
| γ-HCH | + | + | + | + | + | + | + | + |
| Aldrin Dieldrin | + | + | + | + | + | + | + | + |
| Heptachlor, Heptachlor-epoxide | + | + | + | + | + | + | + | + |
| DDT, DDE, TDE-isomers | + | + | + | + | + | + | + | + |
| Chlordane, Oxychlordane | + | + | + | + | + | + | + | + |
| Endrin | + | + | + | + | + | + | + | + |
| Deltaketoendrin | | | | | | + | + | |
| HCB | − | − | + | + | + | + | + | + |

+  applicable
−  not applicable (poor recovery)

For gaschromatography a broad variety of suitable systems is available. Electron capture  detectors have been proven to be most useful for the determination of organochlorine pesticides. Data for repeatability and reproducibility  which should normally be reached are available and given in the IDF-Standard. The difference between the maximum und minimum of three test results must be less than the following values:

| Residue level mg/kg | Difference mg/kg |
|---|---|
| 0.01 | 0.005 |
| 0.1 | 0.025 |
| 1 | 0.125 |

For the reproducibility the following values, based on experimental evidence, should be used:

| Residue level mg/kg | Difference mg/kg |
|---|---|
| 0.01 | 0.01 |
| 0.1 | 0.05 |
| 1 | 0.25 |

Normally the pesticide content of dairy products is expressed on a fat base. However, in the case of milk products with a low fat content it is better to express results on a product base, because the fat content of these low fat dairy products varies widely depending upon which method is used to extract the fat.

It is advisable to determine the <u>fat content</u> by appropriate method and report the result together with the pesticide content. It should be stated how the pesticide content is expressed, in mg/kg on a fat base or on a product base.

For the extraction of fat and pesticides and the fat determination the following extraction methods are used:

- Soxhlet extraction for non-liquid milk products
- Column extraction for all milk products
- AOAC extraction for milk and liquid products

If the extracted fat is used for the determination of pesticides, evaporation of organic solvent solutions should not be allowed to go to complete dryness as this may result in loss of pesticides.

For toxicological evaluation of pesticide residues in milk a comparison of the concentrations, found in milk, and the acceptable daily intakes, which are established by WHO/FAO, should be carried out. In a model calculation, which is valid for numerous countries, one can anticipate that about 1 l of milk is consumed by an adult each day (as fluid milk or milk products). If this milk contains about 4% fat, 35 up to 40 g of milk fat are consumed daily. In the following Table 29 a calculation is given for some pesticides, showing that by the average amount of pesticides in milk only a small portion of the acceptable daily intake is taken[19].

Organophosphates (organic phosphoric acid esters) and carbamates[23].

These compounds have decisive advantage over the organochlorine pesticides. They are generally much less persistent and are degradable to compounds which do not accumulate in animal organisms to a great extent. The comparatively rapid decomposition of these compounds by physico-chemical processes in the environment as well as by enzymatic processes in the animal body prevents the build-up of significant residues in milk.

Most of the existing organophosphates correspond to structures of phosphates, phosphonates, phosphorothionates, phosphorothiolates and phosphorodithioates.

Carbamates are derivatives of esters of carbamic acid ($HO-CO-NH_2$). The mode of action of the carbamates is essentially the same as that of the organophosphates. The acute toxicity is in most cases of the same order or lower than that of the organophosphates.

Milk and milk products can be contaminated with residues of organophosphates and carbamates from various sources:

- Animal feed which has been directly treated with insecticides
- animal feed manufactured from plant material which has been treated during growing season
- use of insecticides directly against parasites on animals
- use of insecticides in animal dwellings
- direct contact of animals with treated areas (grazing land, etc.).

The contamination of milk due to oral intake of organophosphorus or carbamate insecticides is, at the most, very small since most organophosphorus and carbamate compounds are readily metabolised in the alimentary tract and in the organs associated with digestive processes. The hydrolysis by enzymes and enzymatic systems is the major metabolic route of the organophosphates and carbamates.

A considerable amount of work has been done on excretion of residues of organophosphates and carbamates in milk. The results could be summarized in so far that residues are detected in milk only for several hours or very few days after application. In most cases the level of residues is at the limit of detection or in the range of 0.1 to less than 0.01 mg/kg. Only in cases of direct treatment of lactating cows might the residues in the milk of individual animals reach 1.0 mg/kg at the milking immediately after treatment. In commercial milk the residue level, of course, is usually much lower.

## Toxic trace elements ("heavy metals")[23,24,25]

The synthetic organic environmental chemicals and pesticides described above have exerted an influence on the biosphere only during the last few years. With metals and their compounds, in the environment, however, all living beings have been living for millions of years already. Defined trace elements act in small amounts as catalysts, and metal-organic substances in the body have either gained vital importance (iron, iodine, cobalt, manganese, molybdenum, zinc) or have proven to be conducive to vital functions (fluorine, chromium, selenium). As regards other heavy metals as lead, cadmium and mercury (Pb, Cd and Hg) physiological functions are not known and increased concentrations due to environmental factors may have toxic effects.

Toxic heavy metals present in soil or emissions are to be expected also in foods, the concentrations, however, being partly so insignificant that they are detectable only by using time-consuming methods of modern microanalytical technics. The most important analytical instrumentation is atomic absorption spectrophotometry by which lead, cadmium and other trace elements can be detected following appropiate preparation within the microgram range/kg food (ppb).

Unlike chlorinated hydrocarbons which are able to accumulate in the food chain due to their high persistency and lipophilic properties, the inverse phenomenon is observed with respect to toxic trace elements present in emissions from the industry and the traffic as lead, mercury, cadmium etc. In this context, the lactating dairy animal is acting as a partly very effective metabolic filter, which is intercalated between contaminated grazing land and emissions, on the one hand, and man as the milk consumer, on the other hand. Prior to the transfer of lead or cadmium from feed into the milk accumulation of these toxic heavy metals occurs in organs as kidneys and liver and, further, in the skeleton. In both the liver and the kidney there exist mechanisms causing a partly specific binding of heavy metals.

Lead concentrations which can be expected in milk are normally in the range between 0.01-0.03 mg/kg.

Cadmium levels are very low and normally do not exceed 0.002-0.003 mg/kg.

For the determination of the more important toxic trace elements such as lead, cadmium and others atomic absorption spectrophotometry (AAS) is normally used. Depending on the type of product, the following steps have to be mentioned:

- Ashing of the product
- Dry or wet digestion
- Extraction of the metal-complex using e.g. Dithizone (APDC) (optional)
- Determination in the AAS (flameless) with background compensation ($D_2$ or Zeeman)

Under practical conditions there are many factors of influences, which have to be regarded.

## Nitrate, nitrite and nitrosamines[26,27,28]

During the last years scientists have paid increasing attention to nitrate contents in foods of animal and plant origin, for nitrate is reduced to nitrite in foods and the body. Nitrite reacts with haemoglobin and leads to formation of methaemoglobin. It is further possible that N-nitroso compounds are formed in acid medium by reaction of secondary amines with nitrite ions. These have been shown to be carcinogenic, mutagenic and teratogenic.

Milk and milk products are apparently of negligible importance as regards the nitrate/nitrite content in the human body. However, one has to bear in mind that in numerous countries addition of 15-20 g nitrate/100 l cheese milk is permitted in the manufacture of certain cheese varieties (semi-hard cheese).

The daily admissible intake of nitrate-N is 48 mg (ADI). Average consumption of milk incl. milk products contributes only 2 o/oo to this quantity.

Nitrite is very active and may lead to nitrosation reactions in foods and also in the digestive tract following consumption of foods. The resulting nitrosamines exhibit the following basic structure

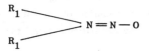

Although in forage for lactating dairy animals formation of nitrosamines is possible and has also been detected in silage as a function of dry matter content, nitrosamines have not been detectable in the rumen content of cattle fed with silage. Both passage and formation in milk is practically excluded to a far extent.

Extremely high and partly toxic doses of nitrosamines administered orally to lactating cows may lead to passage of nitrosamines into milk, the proportion being below 1%. Under practical conditions nitrosamines have not been detected to date in cow's milk.

In order to assess properly the whole problem of nitrate, nitrite and nitrosamine contents attention has to be given to the fact that the proportion of nitrate resulting from milk accounts for less than 1 % of the total nitrate intake of man, whilst that from foods of plant origin, incl. cereal products and bread and confectionary accounts for 70-80% and that from meat and meat products 10-20%.

For the <u>determination of nitrate and nitrite</u> several measuring principles can be used:

1. directly with ion-selective eletrodes (nitrate)

2. indirectly by measuring derivatives of nitrate and nitrite.
   For this group the following principles should be mentioned:
   - nitration of aromatic compounds with nitrate
   - forming dyes by azo-coupling reaction with nitrite

The latter method has the advantage that it can be carried out by means of continous flow analysis with the Auto Analyzer II System. The principle is the following:
First nitrate is reduced to nitrite since only with nitrite an azo-coupling reaction can be done. This reduction is carried out with a hydrazine solution and is catalysed by copper ions. It should be noted here that in other methods this reduction is often performed with cadmium.
Secondly an azo-coupling reaction with nitrite, sulphanilamide and N-(1-naphthyl)ethylenediamine dihydrochloride is performed.
Finally the extinction of the dye is measured at 520 nm.

For the determination of nitrate and nitrite in one sample the analysis has to be done twice. First without hydrazine solution (determination of nitrite) and secondly with hydrazine solution (determination of the total amount of nitrate and nitrite). The difference of both determinations gives the content of nitrate.

# Mycotoxins[29-33]

Mycotoxins are formed by defined moulds. At present about 130 mycotoxins are known as secondary metabolites of cell metabolism. Formation of mycotoxins takes generally place where defined environmental conditions (temperature, humidity etc.) are prevailing and the biochemical reactions for maintaining the processes of life have either come to an end or may continue unhindered. Hence it follows that mycotoxins are formed by potentially toxigenic fungi only under specific conditions. The toxin-producing mould itself is either not influenced by the formed mycotoxins or to a slight degree only.

The main mycotoxins, their producers and their effects on mammals can be seen in Table 11. Aflatoxins are the best known mycotoxins.

Table 11: Mycotoxins, their producers and effects on mammals

| Toxin | Producers | Effects on mammals |
|---|---|---|
| Cumarin derivates | | |
| Aflatoxins | Aspergillus flavus, A.parasiticus, Penicillium puberulum | Carcinogenic, hepatotoxic |
| Ochratoxine | A.ochraceus, A.melleus, P.viridicatum | Hepato- and nephrotoxic |
| Anthrachinone | | |
| Islandicin | P.islandicum | Hepatotoxic |
| Luteoskyrin | P.islandicum | Carcinogenic, hepatotoxic |
| Pyrones | | |
| Patulin | P.expansum, P.patulum, P.urticae, A.clavatus, Byssochlamys fulva, B.nivea | Carcinogenic, general Cellular poison |
| Citreoviridin | P.citreoviride | Nephrotoxic |
| Kojic acid | A.oryzae, A.flavus, A.candidus, A.nidulans, A.tamari | Neurotoxic |
| Xanthones | | |
| Sterigmatocystin | A.versicolor, A. nidulans | Carcinogenic |
| Polypeptides | | |
| Islanditoxin | P.islandicum | Hepatotoxic |
| Macrolides | | |
| Zearalenon | Fusarium roseum var. graminearum | Estrogen, abortions, Sterility |
| Steroids | | |
| Trichothecene | F.tricinctum | Dermotoxic, enterotoxic |
| Sporofusarin | F.nivale | Hemotoxic, |
| Poafusarin | F.poae, F.lateritium | Alimentary toxic aleukia |
| Ergotalkaloids (Peptidester of lysergic acid) | Claviceps purpurea | Neurotoxic |

Aflatoxin $B_1$ is considered the prototype of hepatic poison responsible for tumor growth. Minimal and maximal growth temperatures of Aspergillus flavus are influenced by humidity, oxygen concentration, type of nutrient medium etc.

Experiments with animals have shown that aflatoxins present in feed given to lactating dairy animals can pass into the milk (1-3 %). This may be considered to be practically the only source of contamination. Aflatoxin is eliminated in the form of aflatoxin $M_1$ (milk toxin), aflatoxin $B_1$ being converted by mitochondrial oxidases in the cell of the liver.

Main carriers of aflatoxin $B_1$ are the feed components peanut meal and cotton seed. It is possible, by excluding these components from supplementary feed for lactating animals to keep aflatoxin $B_1$ residues in concentrates within the range of few µg/kg and to produce a milk containing lowest quantities of aflatoxin $M_1$ (10-100 ng/kg).

For the detection of lowest quantities of aflatoxin $B_1$ and $M_1$ in feed and in milk effective analytical methods are available. Clean-up of sample extracts plays a decisive role. According the latest state of knowledge it can be performed using gel permeation chromatography. Detection of aflatoxins is performed by thin layer chromatography . The detection limits in milk are as much as 1 ng/kg aflatoxin $M_1$; in feed residue concentration of 1 µg aflatoxin $B_1$/kg (dry matter) can easily be determined[34].

There is almost no information available on carry-over of further mycotoxins from feed into milk. This will form an important part of future research.

## Polychlorinated biphenyls (PCBs)[35,28,36]

The physical and chemical characteristics of PCBs, such as their thermal stability, resistance to oxidation, acids, bases and other chemical agents, and their insolubility in water, make them very important in environmental studies. As to the high degree of stability, the gradual build-up of PCB levels in the environment can be expected. Theoretically, a total of 210 different polychlorinated biphenyls could exist.

Polychlorinated biphenyls are marketed in various countries under different trade names (Aroclor, Clophen, Phenochlor, Kanechlor, and others) and used industrially in synthetic resins, synthetic and natural rubber paint, varnish, wax, plasticizers and extenders, hydraulic fluids, extreme pressure lubricants, heat transfer media and in electronic and electrical components. The use of PCBs in pesticide formulations to improve effectiveness has also been suggested.

PCBs have been found as residues in milk and milk products in many countries of the world (North America, Europe). Their presence in milk has been traced to the accidental contamination of cattle feed. In the United States, a coating containing PCBs used on the inside of farm silos led to the contamination of dairy products. In milk of other countries rather low values are reported. These suggest a more indirect source of contamination.

The gaschromatographic pattern of PCB residues in milk samples indicates that the lower chlorinated peaks in technical PCB formulations were reduced. The absence or reduction of the lower chlorinated peaks suggest that they are selectively metabolized, absorbed or excreted by the lactating cow. The principle route of excretion of unmetabolized PCBs in the lactating cow is via the milk. The metabolites are excreted in milk and urine. Feeding trials of dairy cows showed that the individual cows excreted about 10-20 % of the daily consumed dose via the milk. This figure is very similar to that of many chlorinated insecticides[19].

As the physical and chemical properties of PCBs are similar to that of the DDT family, the analytical behaviour of these two groups is similar. Both classes of compounds are highly lipophilic, so that accumulation of residues takes place in the fat deposits of animals. Therefore, as a rule, partitioning with organic solvents and absorption column clean-up is required after initial extraction from the fat. Sophisticated gaschromatographic techniques, such as temperature programming and high resolution columns with defined carrier materials, allow the separation of the different PCB isomers and their metabolites. For the identification of different PCBs the capillary gaschromatographic technique is recommended. This method is especially useful for the combined determination of pesticides and PCBs.

The Group E12[37] of the International Dairy Federation has worked out a first draft for recommended methods for the determination of PCBs in milk and milk products. Taking into consideration the different technical equipment available in the laboratories, this draft contains the following methods:

- Gaschromatographic determination of PCBs (and organochlorine pesticides) on packed columns
- Quantitative determination of significant individual chlorobiphenyls by splitless glas capillary gas chromatography
- Other methods of determination (perchlorination, dechlorination)

The extraction of fat and the clean-up of the fat extract is performed in the same way as in the case of the other chlorinated hydrocarbons.

Information on the presence of PCBs in milk is very limited. Values between 0.01 and more than 1.0 mg/kg (fat basis) are reported. In Europe residues of the order of magnitude of 0.2 to 0.3 mg/kg butter fat can be expected.

The significance of polychlorinated biphenyls in milk seems to be limited, as levels of PCBs in milk in the European and North American literature are reported to be generally low and the contamination in some countries is not as frequent as in the case of organochlorine pesticide residues.

Detergents and disinfectants[38]

Detergents and disinfectant residues can contaminate milk during or after milking. The concentrations of the different components usually present in working solutions of disinfectants and detergents are given in Table 12.

Concentrations of alkali and acids may sometimes be higher, when used for certain cleaning purposes under certain conditions.

Contamination may also occur from residues on the teat skin. Here iodine, quaternary ammonium compounds, chlorhexidine and chlorine have to be mentioned. Residues remaining on teat skin and which gain access to the milk will be of little concern, if good practices are followed. The use of quaternary ammonium compounds would be the least desirable for this application due to the tendency to adhere to skin and their bacteriostatic effects in milk used for fermentation processes.

Provided that equipment surfaces are subjected to an adequate clean water rinse subsequent to cleaning and sanitizing, there should be no significant residues gaining access to the milk from this source. The adsorptive nature of quaternary compounds make them difficult to rinse completely from surfaces, and careless rinsing may lead to significant amounts finding their way into the milk. The use of iodophors for equipment which contains rubber may lead to adsorption of iodine by the rubber components.

The toxicological evaluation of residues of the different detergents and disinfectants shows that under conditions of good practice hazards to human health could not be expected. There is substantial evidence in the literature to demonstrate that when the appropriate cleaning and sanitizing product is used as directed, there is very little likelihood of significant quantities of residues in the milk.

Table 12: Concentration (%) of components in working solutions for cleaning and sanitizing

| Product | Active alkalinity | Acid | Sequestrant | Surfactant | Emollient ammonium compound | Iodine | Chlorine | Quaternary Q.A.C. | Chlorhexidine gluconate |
|---|---|---|---|---|---|---|---|---|---|
| Alkaline detergent | 0.02–0.5 | | 0.005–0.20 | 0.0–0.05 | | | | | |
| Alkaline chlorine det-sterilizer | 0.02–0.5 | | 0.005–0.10 | 0.0–0.05 | | | 0.015–0.05 | | |
| Alkaline quaternary det-sterilizer | 0.02–0.5 | | 0.005–0.10 | 0.0–0.05 | | | | 0.015–0.025 | |
| Acid detergent | | 0.02–0.5 | | 0.0–0.05 | | | | | |
| Amphoteric surfactant | | | | 0.1–0.20 | | | | | |
| Acid iodophor det-sterilizer | | 0.02–0.06 | | 0.04–0.10 | | 0.0025–0.07 | | | |
| Acid quaternary det-sterilizer | | 0.02–0.06 | | 0.02–0.05 | | | | 0.015–0.025 | |
| Acid anionic sterilizer | | 0.05–0.10 | | 0.02–0.05 | | | | | |
| Iodophor teat dip | | | | 0.10–2.00 | 1–10.0 | 0.5–1.0 | | | |
| Chlorhexidine teat dip | | | | | 1–10.0 | | | | 0.15–0.3 |
| Chlorine teat dip | | | | | | | 2–4 | | |
| Quaternary teat dip | | | | | 1–10.0 | | | 0.2–0.5 | |

det = detergent

234

## Literature

1   W  Heeschen, Deutsche Tierärztliche Wochenschrift, 1980, 87, 209.

2   International Dairy Federation (IDF), 1979, Bulletin, Document 113.

3   Food and Agriculture Organisation/World Health Organisation (FAO/WHO), Report of the Codex Committee on Pesticide Residues, 1982.

4   J. Hamann, A. Tolle and W. Heeschen, IDF-Bulletin, Document 113, 1979, 43.

5   H. Knothe, Münchener Medizinische Wochenschrift, 1963, 105, 173.

6   W. Heeschen, IDF-Bulletin, Document 113, 1979, 30.

7   G. Rosenberger (Ed.), "Krankheiten des Rindes", Verlag Paul Parey, Berlin-Hamburg, 1978.

8   W. Heeschen, IDF-Bulletin, Document 113, 1979.

9   M.J. Kaiser, Thesis, Universität München, 1977.

10  W. Heeschen, IDF, A-Doc 46, 1979.

11  B. Marek, IDF-Bulletin, Document 113, 1979, 17.

12  D.G. Dunsmore, S.M. Wheeler and R.N. Barnes, Aust.J.Dairy Technol., 1977, 32, 119.

13  D.G. Dunsmore and S.M. Wheeler, Aust.J.Dairy Technol., 1977, 32, 166.

14  International Dairy Federation (IDF), Document 85, 1975.

15  E. Schumacher, Wiener Tierärztliche Monatsschrift, 1975, 62, 221.

16  International Dairy Federation (IDF), Document 152, 1983.

17  W. Heeschen, Milchpraxis, 1979, 17, 42.

18  H.F. Smyth and J.C. Calandra, Toxicol. and appl. Pharmacol., 1969, 14 315.

19  J.T. Snelson and L.G.M.Th. Tuinstra, IDF-Bulletin, Document 113, 1979, 6.

20  International Dairy Federation (IDF), Bulletin, Document 113, 1979.

21  W. Heeschen, Tierärztliche Umschau, 1982, 37, 538.

22  International Dairy Federation (IDF), Standard 75A:1980.

23  International Dairy Federation (IDF), Document 105, 1978.

24  C. Reilly, "Metal Contamination of Food", Applied Science Publ., London, 1980.

25  F.W. Oehme, "Toxicity of Heavy Metals in the Environment", Marcel Dekker, New York, 1978/1979, Part 1 and 2.

26   H. Nijhuis, W Heeschen, A. Blüthgen and A. Tolle, Milchwissenschaft, 1980, 35, 678.

27   A. Askar, "Fortschritte in der Lebensmittelwissenschaft", Technische Universität, Berlin, 1976, Vol. 4.

28   World Health Organisation (WHO), Environmental Health Criteria, Geneva 1976, Part 2

29   I.V. Rodricks, Advances in Chemistry Series, American Chemical Society, Washington D.C., 1976.

30   I.V. Rodricks, C.W. Hesseltine and M.A. Mehlman, "Mycotoxins in Human and Animal Health", Pathotox Publ., Park Forest, South Illinois, 1977.

31   Hibbet and Heathcot, "Aflatoxin", Elsevier, Amsterdam-London, 1980.

32   F. Kiermeier, "Mykotoxine in Lebensmitteln", Gustav Fischer Verlag, Stuttgart-New York, 1981, p. 245.

33   I. Reiss (Ed.), "Mykotoxine in Lebensmitteln", Gustav Fischer Verlag, Stuttgart-New York, 1981.

34   H. Nijhuis, A. Blüthgen and W. Heeschen, Milchwissenschaft, 1982, 37, 449.

35   V.W. Kadis, IDF-Bulletin, Document 113, 1979, 63.

36   International Agency for Research in Cancer (IARC), Monographs, Lyon 1978, Vol. 18.

37   International Dairy Federation (IDF), Appendix to Questionnaire 2483/E, 1982.

38   P. Mannaert, IDF-Bulletin, Document 113, 1979, 57.

# Artefacts in Milk Products

By H. Klostermeyer

MUNICH INSTITUTE OF TECHNOLOGY, FREISING-WEIHENSTEPHAN, D-8050 FREISING,
WEST GERMANY

There is no doubt, milk is a complex system of many different components.
Already during storage and especially during processing there do not only
occur the intended effects but also changes not asked for in the chemical
composition and in the physical structure. It is impossible to report comprehensively upon all the reactions in one paper, hence it is necessary to
make a choice. In the following lecture especially processes are discussed
which the author thought to be particularly interesting and to the knowledge
of which he could contribute.

A chemist should review the reactions of interest systematically according
to the underlying mechanisms - on the one hand radical reactions and on
the other hand reactions including ion mechanisms because both show common
properties in regard to the causes as influence of light and oxygen on the
milk or in regard to the polarity, the ionic strength, the pH of the medium
etc. Unfortunately it is sometimes difficult to keep to this classification
consequently because in complex reactions as for example lipid oxidation
or Maillard reaction both types of reaction are combined. Therefore the
traditional classes of substances are chosen as a clue for the subject even
if it is not always possible to keep to it.

Gases

The modern technologies of milking, transport, storage and pumping cause
an intensive aeration of the milk. The dissolved gases reduce the freezing
point; that complicates for instance the detection of watering of milk.
Dissolved carbon dioxide causes changes in acidity while dissolved nitrogen
has no influence. Oxygen plays an important role in many different reactions;
some are mentioned below. As processed milk usually contains lactic acid
bacteria which need a low redox potential oxygen is generally an undesirable
contaminant of milk. The oxygen content of milk can be easily measured by
electrochemical procedures.

## Minerals

Since the forties it is known that already cooling (1) and especially heating
of milk influence the thermodynamically labile state of many milk salts
and thus for example change the dialysability of $Ca^{2+}$, phosphate and other
ions. Using ion-selective electrodes numerous scientists have recently tried
to determine to what extent for example $Ca^{2+}$ is dissolved ionogenically
or is undissociably bound to colloidal or even crystalline phosphate etc.
The results of these studies were often inconsistent. A possible explanation
is that the investigated system is more labile than most workers would assume.

Too little work is done about the behaviour of the trace elements. Since
most of the nutritionally important cations of milk form difficultly soluble
phosphates it can be expected that by the crystallisation of phosphates
(e. g. during heating of milk) they become insoluble and therefore physiologi-
cally unavailable. In the new literature there are indications of such effects
(2). We ourselves (3) made some investigations on this phenomenon and found
out that several trace elements in high heat milk were about 10-15 % less
dialysable - whether this is of importance for the biological availability
of these elements is not cleared up. To our opinion in this case for unambiguous
results it is required to work with isotope-labeled milks.

The progressing crystallisation of phosphates in the intersubmicellar regions
of the casein micelles can be measured by X-ray methods, besides the crystals
formed can be identified (4). Most of the crystals seem  to be tertiary
phosphates. According to the $pK_3$-value of phosphoric acid these crystals
will react strongly basic on their surface - so they are (because of their
fine dispersion and their immense surface) ideal catalysts for $OH^-$-catalyzed
reactions during heating of milk.

An example for this catalysis is the formation of lactulose but also the
anomalous dependence on temperature of the activation energy for the formation
of lysinoalanine in milk indicates the presence of a catalyst; reaction
increases with the heat treatment and the formation of "milk stone" deposits
in heating equipment (6). Milk heaters in which sediments are built during
the working-period show also increasing concentrations of lactulose during
the working-period although the heat-treatment remains the same (7). In
practice, it has to be distinguished between deposits from preheated milk
consisting mainly of minerals and deposits from not heated milk including
many organic materials. The crystallization of tertiary phosphates leads

automatically to an acidification of the milk; the measurable change of
the pH can be used to evaluate the heat treatment and to distinguish between
sterilized milk and UHT milk (8). Besides, the titration behaviour of the
casein micelles and hence the buffering capacity are clearly different according
to the heat treatment (9).


## Lipids


In the field of lipids during the last years the most knowledge was obtained
of the lipid oxidation or better of the oxidation of unsaturated acyllipids
(10). The complicated reaction coil is disentangled by experiments with
model systems. This would hardly be possible without the HPLC-technology
which allows to separate and determine even labile intermediates, *i. e.*
hydroperoxides and peroxides.

Since the late thirties it is known that the autoxidation of unsaturated
acyllipids is a radical chain reaction. This reaction mainly can be started
by the following four factors:
- Oxidation with oxygen to hydroperoxides in presence of lipoxygenases
- Photooxidation in presence of a potent sensitiser like riboflavin
- Reaction with oxygen to peroxides in presence of catalytic amounts of
  $NO_2$
- Reaction with oxygen to peroxides in presence of catalytic amounts of
  $NO_2$ or $SO_2$

Certain heavy metal ions as well as hem and hem proteins and light play
a keyrole in splitting the hydroperoxides to start the reaction chain.
Undoubtedly many not exactly defined "off-flavours" of milk and milk products
could be prevented if it was possible to eliminate all these reactions.

For the analyst it is a remarkable new result that only fatty acids with
three or more double bonds can form those bicycloendo-peroxides which under
conditions of the thiobarbituric acid test break down to malondialdehyde
and can then be determined (11). Since butterfat contains only small but
varying amounts of these fatty acids it becomes clear why the thiobarbituric
acid test often does not correlate with the sensory deterioration - this
observation we made too in our laboratory during long term storage experiments
with butteroil (12).

As the intermediates of the lipid oxidation are highly reactive substances
they take part in many undesirable reactions in foods. It is also known
that they play an essential role in pathogenous biological processes. We
can assume that in future much more important facts about this subject will
be published.

In the recent years apparently less work was done about the hydrolytic rancidity
of milk - sponaneous or induced - than ten years ago (13). While in the
meantime simple tests for the determination of free fatty acids and hence
of the extent of lipolysis became available there still seems to be a lack
of a universal, simple, sensitive method for the determination of the lipases
causing the lipolysis. According to the facts known about the mechanism of
enzymatic lipolysis (dependent on activators as well as on size and nature
of the lipid/water contact surface) it will be difficult to make progress
with this subject because a standardization of the experimental conditions
is extremly complicated. But as long as it is not possible to detect even
smallest activities of lipases correctly it will be  difficult to recognize
and to define nonenzymatic lipolytic processes as they are used in the technical
fat saponification and which the chemist could expect in case of such highly
emulsified triglycerides like milk fat.

The consequent use of HPLC now allows a determination of tri- and partialgly-
cerides which is simpler than gas chromatography (14); this will certainly
offer interesting analytical work during the next years. Nowadays strongly
polar lipids - particularly the phospholipids - can be separated and determined
quantitatively by a rather simple method (15). Among other things this offers
a better chance to watch changes of these substances which are thought to
be main causes for the sensory deterioration of products like butter milk.
In this connexion it should be mentioned that the enzymatic analysis presents
still unutilized potentials - according to the previous experiences, however,
the sensitivity of these methods for the determination of phospholipids
in milk products needs to be increased.

Only a few facts are known about the protein/lipid interactions because the
investigation of these reactions is very complicated. Meanwhile it was found
that for example heating of fat/protein systems causes protein cross-linking
and "damage" of amino acids (e. g. lysine, methionine, tryptophan) - whatever
this means. Ten years ago workers were of the opinion that these reactions
only occur at temperatures of more than 150 °C - these temperatures are
not used in the milk industry. Further experiments showed that this opinion

had to be revised: in the meantime a temperature limit of about 50 °C is discussed (10) - in my view, the new opinion is a result of better analytical possibilities.

On principle a reaction like for example the esteraminolysis should already happen at lower temperatures. At normal temperatures the turn-over rates in a system like milk are to be extremely small because of the actually very low concentrations of the reaction partners in solution. But it is certainly only a matter of time and analytical progress to obtain more information in this field.

The analysts now have started to work on the changes of the fat globule membrane - a subject of immense importance for the practice. These changes already start during cooling of the raw milk. The resulting shifting of the native membrane (in favour of the caseines) can for example easily be demonstrated by the migration of the xanthinoxidase into the fat-free phase (16). During the crystallization of the lipids  the fat globules are deformed so that part of the lipids, the so-called "free" fat can be extracted by solvents. For its determination several new methods have been developed; they detect different amounts of "free" fat because this is only defined by the method used but all these methods allow technologically relevant statements about the "damage to fat" (17).

Since the lipids in native milk of 38 °C (cows-body heat) are liquid their crystallization has - strictly speaking - to be regarded as artefact, too. For measuring the extent of the crystallization the NMR spectroscopy has proved useful (18). Nowadays melting and freezing curves of butterfat provide exact data for the optimal course of the so-called physical cream ripening leading to butter of an optimal spreadability. The transformation of the curves can be easily done by a minicomputer; the resulting curves can be interpreted very well (19).

## Carbohydrates

In contrast to other milks the carbohydrate composition of cows' milk is rather simple; the main constituent is lactose. Additionally milk contains several sugars glycosidically bound to proteins.

On principle it should be possible to transform lactose into several other sugars in aqueous solutions by catalysis of $H^+$- and especially $OH^-$-ions; practically this happens in milk only at higher temperatures whereat lactulose and epilactose are the most important reaction products (20). At given pH values the formation of lactulose in milk, cream etc. is proportional to the heat treatment of the product, i. e. the lactulose content is an analytical criterion for the characterization of the heat treatment (21). Significantly higher and during the working-period increasing amounts of lactulose, however, are observed when milk is subjected to indirect UHT treatment. In this case the milk is heated first to precipitate the whey proteins onto the casein micelles and then the UHT treatment follows. In the milk heaters a nearly pure mineral and strongly basic (molar ratio Ca:P = 1,5 : 1) deposit is formed which is in contact with the milk for quite a long time so that the formation of lactulose is catalyzed. The catalytic effect of fouling deposits is for example proved by the fact that a 5 % aqueous lactose solution heated for 20 min at 110 °C contains 20 mg lactulose/l; when some mg of ground deposit are added to the lactose solution before heating the amount of lactulose formed is about 200-300 times higher (7).The result in milk is similar. We have to start from the principle that at least all alkali catalyzed reactions are accelerated in equipment containing fouling deposits.

For the specific determination of smallest amounts of lactulose in presence of high amounts of lactose we developed an enzymatic procedure (22) while other workers announced an effective HPLC method (23). HPLC and enzymatic methods are now the most important techniques for the routine analysis of sugars in milk. In the next years the coupling of gas chromatography with mass spectrometry and computer analysis is to increase our knowledge of artefacts in milk (products) significantly. Investigating a methanolic extract of lyophilized heated milk one of my Ph.-D. students identified on the first attempt Maillard compounds as yet unknown in milk (24) - how efficient this technique will be when it is used well-aimed!

It would lead too far to report comprehensively on the analysis and products of the Maillard reaction in this paper; I only want to draw the attention to a side reaction. We found smallest amounts of formaldehyde in foods by determining its reaction products with lysine - i. e. $N^\varepsilon$-methyl- and $N^\varepsilon,N^\varepsilon$-dimethyl-lysine - and always found blank values in carbohydrate containing products like milk powder (25). The formaldehyde apparently did not result from the retroaldol reaction of carbohydrates in model systems; it was also formed from desoxylactulosyl-lysine, the so-called Amadori product of lactose (26).

Since we have been able to detect again a $C_5$-fragment of carbohydrates too, namely furfuralcohol (24), it is our working hypothesis at the moment, that the $C_1$- and $C_5$-fragments result from the degradation of the galactosyl residue of desoxylactulosyl amines. So far as we know, the fate of the linked carbohydrate moiety in the course of Maillard reaction of disaccharides is yet not studied.

In this connexion it is remarkable, too, that formaldehyde is a methylating agent - probably a second molecule acts as reducing agent. Already in the well known text-book of Jennes-Patton from 1959 the occurrence of formic acid in heated milk is mentioned - perhaps it is derived from this reaction too.

Since many years because of their rather simple analysis characteristic Maillard components have been discussed as criteria for the heat treatment and the changes of milk during storage (27). Recently the kinetics of their formation have been studied more intensively (28). In heated milk we found a substance from the Maillard complex that can be detected electrochemically (29); perhaps it is possible to determine it continuously thus allowing the control of milk heaters. To our knowledge their is as yet no analytical parameter for this purpose. Unfortunately it was not possible until now to identify this substance.

Perhaps the research in the field of sialic acids should be increased, too. Sialic acids are components of many glycoproteins; in milk they are found for example in the   caseins. In the recent years several methods for the determination of whey in milk powders etc. were developed based on the distri- bution of sialic acid between proteins and peptides (30, 31, 32). In this connexion it should be taken into account that sialic acids generally are rather labile in aqueous solutions (33). Already during heating they are more or less decomposed (34, 35), at basic pH leading to reverse aldoladdition, at acid pH leading to the formation of ammonia and humic acids; sialic acid bound to protein is destroyed to a higher degree than free sialic acid. From these reactions probably interesting analytical parameters can be derived. The sialic acids show a characteristic behavior in affinity chromatography, which might be useful for analytical work too (36).

Among the carbohydrates that occur in milk only in traces the ascorbic acid certainly received most attention. Generally an analytical method was used that was more characteristic for its redox potential than for ascorbic acid

itself (reduction of Tillmann's reagent). On principle this also applies
to the electrochemical    procedures. In the meantime a specific enzymatic
determination has been developed that is now available for routine analysis
(37).

Ascorbic acid and its oxidation product dehydroascorbic acid are oxidized
in milk rather quickly; heavy metal ions act as catalysts. This process
is of importance for the practice; recently it has been intensively studied
in UHT milk because the competitive oxygen consumption of the ascorbic acid
oxidation retards the disappearance of the cooked flavour (29). Since ascorbic
acid was found to be a strong inhibitor of the alkaline phosphatase the
result of the phosphatase test - used as indicator for the heat treatment
of milk - also depends on the content of ascorbate (38).

NPN components

The occurrence of non-protein nitrogen components in milk was intensively
investigated partly using new analytical methods (39). With exception of
urea the content of the other compounds showed only negligible variation
(40).Urea also seems to be the only NPN component the content of which influences the chemical-physical properties of the milk. Fox et al. (41) studied
this phenomenon and found effects on the heat stability, the Maillard reaction
and the buffering capacity of milk.

As every protein chemist knows in buffers containing urea the partial reversion
of the Wöhler reaction leads to the formation of cyanate that can easily
modify proteins (at the amino and the SH-groups); whether this is also important
in milk and not only in model solutions has to be cleared by further investigations (42). To my knowledge it can hardly be estimated at present to what
extent changes in the content of the NPN components - especially of the
orotic acid, the freee amino acids and peptides - actually influence the
growth of the bacteria used in the dairy industry.

Proteins

According to their size and their huge number of potential reactive arrangements in each molecule proteins are the chemically most interesting components
in the milk system - surely they are also the most difficult substances
from the analytical point of view if you want to attribute chemical-physical

changes to certain positions in the molecule. Hence most of our knowledge
for example of "the SH-groups" or "the available lysine" is rather global
up to now without determination of individual reaction sites.

It would already be a great progress if all workers tried to differentiate
between reactions without changes in the primary structure of molecules
and those with real chemical reactions, i. e. with changes of covalent bonds.
The worst example of this problem seems to be the vague use of the term
"denaturation". In the field of biophysics this exclusively means processes
without changes in the primary structure of a protein; this is in good agreement
with the international chemical and biochemical nomenclature committees
which classify every molecule with a different primary structure as a chemical
individual with its own appellation.

It was to be admitted that it is difficult or even undesirable to denominate
derived proteins before the structural changes are known and thus a basis
for the nomenclature is given. Nevertheless in cases where more facts were
known a compromise with the nomenclature was found. As an example the investigations about the denaturation of ß-lactoglobulin can be mentioned in which
several degrees of denaturation had to be distinguished (43).

Similar extensive studies of the denaturation of other milk proteins were
not made yet although the denaturation is of great importance for the reactivity
of the globular milk proteins and for the technological properties of milk.
Because of analytical difficulties  experiments with systems of protein
mixtures are nearly impossible up to now. We are content that using rather
simple measuring systems like the differential thermoanalysis we can get
technologically relevant information  about those temperature ranges where
the denaturation of the several proteins occurs; besides it is possible
to detect influences of the medium at reasonable costs (44, 45).

For a complex system like milk the HPLC certainly will be useful with stationary
phases that separate proteins not only according to the molecular weight
(45)but also according to the molecular structure and perhaps to the changed
reactivity. In the field of milk protein denaturation very expensive electrophoretic methods were used; in my opinion the success was only limited;
I do not exclude our own expensive studies with antibodies against different
derivatives of milk proteins (46).

While heat-induced changes in the system of milk proteins were noticed early
and hence studied the effects of milk cooling were discovered rather late.
Indeed from the changes in the dialysability of milk salts after cooling
indirectly conclusions could be drawn on changes of the casein micelle but
intensive studies on the process in the protein system were not possible
because the analytical methods for the individual milk proteins were not
available yet. On the one hand these processes can be watched indirectly
- for example the migration of the original fat globule membrane proteins
during cooling of the milk can be detected by the distribution of enzymatic
activities between cream and skim milk; on the other hand the reoccupation
of the free membrane sites by casein (sub)micelles can be shown directly
by electronmicroscopy (47). The electron microscopy has also proved efficient
for many other studies on the protein system of milk.

The proteolysis of milk proteins - especially of the non-globular and hence
easily digestible caseins - belongs to the undesirable changes that occur
particularly after cooling of milk. The responsible enzymes are plasmin
(an enzyme of the milk) and exogenous proteinases of microorganisms, mainly
of Ps. fluorescens. It is still unclear whether the thrombine now detected
in milk by a very sensitive method (48) is also of importance  for the practice.

Because of their economic importance proteolytic processes have been studied
intensively in the recent years. Detection methods using synthetic peptide
substrates (48 - 52) or dye-marked collagen (53) and casein have been elaborated
(54). It has to be regarded that the proteinase activities measured by means
of these substrates are not absolutely parallel to the actual decomposition
of proteins in milk (54). According to our own experiences an increase of
the detection sensitivity by another power of ten is desirable.

For the investigation of changes in the protein system of milk caused by
proteolysis electrophoretic methods have proved useful in numerous cases;
the extremely hydrophobic $\gamma$-caseins formed from ß-caseins by hydrolysis
with plasmin can even be extracted by organic solvents (55) for analysis.
Since the undesired proteolytic changes in milk occur only at a few positions
in the milk proteins the fission fragments are exactly known in some cases
(56).

Several changes in the milk protein system affect exclusively or especially
certain amino acids; the analysis is based on these amino acids. The amino
acids cysteine/cystine are of special importance. They are very reactive
and can be detected by many methods. Most experiences we ourselves got with

polarographic methods for the determination of cysteine/cystin (57, 58).
Other workers use for example colorimetric methods of amino acid analysis
(59, 60).

Unfortunately these analytical techniques are not applicable for the most
important reaction of proteins in the dairy industry, i. e. the thiol-disulphide
exchange; in this case only methods can be used that detect changes in the
molecular weight, e. g. the gel permeation chromatography and electrophoresis
techniques in gels including SDS-electrophoresis. Unfortunately even this
method is insufficient to characterize the high molecular products of the
thiol-disulphide-exchange that are formed during many processes in the dairy
industry, because of their insolubility in all suitable buffers.

Even the often cited crosslinking products of ß-lactoglobulin and ϰ-casein
are not characterized up to now. Investigations with isotope-marked proteins
suggest that all cysteine/cystine proteins in milk are included in the polymeri-
sation according to their quantitative relation; this has been proved at
least for α-lactalbumin besides ß-lactoglobulin and ϰ-casein (61). In case
of larger particles the aggregation of proteins can also be studied by means
of an electron microscope (62), partly even by electronic particle counters
(63); besides the determination of the viscosity and the behaviour at membranes
give further informations.

The heat- and alkali-catalyzed desulfuration of cysteine and cystine residues
has a key position for the formation of the cooked flavour; apparently the
α-lactalbumin plays an important role in this reaction. Sulphur balances
as well as analysis of the resulting products show that α-lactalbumin –
like the structurally similar lysozyme – is surprisingly reactive in an
acidic medium while ß-lactoglobulin behaves as it is expected from other
proteins (64, 65). Without the particular properties of the α-lactalbumin
the formation of the cooked flavour at the pH values of milk could hardly
be explained.

The desulphuration is mainly a ß-elimination at cysteine and cystine leading
to dehydroalanine residues in the proteins. These artefacts are rather stable;
they can be determined photometrically (66) or as pyruvate after hydrolysis
(67). Dehydroalanine is also easily formed from phosphoserine residues and
in certain cricumstances – at least by suitable influence of the adjacent
groups – from serine residues (68).

It is not known yet to what extent serine (D,L-serine!) is rebuilt from
dehydroalanine by addition of water because the analysis is quite difficult
but apparently progress is being made in this field (69). The D-serine strongly
reduces the digestibility of the proteins; thus this phenomenon becomes
relevant for the practice. The racemization of amino acids in proteins and
the resulting losses in the nutritional value caused by the technological
processing are at present still unsolved analytical problems. These phenomena
seems to be also a key for the often noticed differences between biological
and chemical determinations of nutritional damage to proteins (70).

More important than the reaction of dehydroalanine with water is according
to present knowledge the addition of strong nucleophiles, namely of thiol
and amino groups. The addition of cysteine residues leads to chain cross-
linkings with lanthionine residues (67, 71). These can be determined in
hydrolysates using an amino acid analyzer with special separation conditions
whereat double peaks can occur because of different diastereoisomers (72).
The sulphones and sulphoxides of lanthionine have not been studied intensively
so far because they - like those of methionine - occur in originaly samples
only in unimportant amounts - but this theory might be wrong.

Much attention was drawn to the addition of lysine residues to dehydroalanine
leading to the formation of cross-linkings by lysinoalanine, the so-called
LAL; this amino acid is very toxic for rats (73). The most frequently used
determination method for LAL is the ion exchange chromatography after acid
hydrolysis (74); using fluorescent reagents the detection sensitivity is
increased up to the range of ppb (75). This allowed the critical control
of many milk products and led at least in some countries to a more careful
processing of milk. In my opinion the LAL contents of milk products should
less be seen from the viewpoint of a not proved toxicity for man but more
as an indicator for often unnecessary technological treatment of the product.

By addition of ornithine and ammonia to dehydroalanine residues ornithinoalanine
and ß-aminoalanine, respectively, are formed. These amino acids are toxic
for rats, too, but they apparently are not of importance in milk products
(68,75). Because of its long retention time and difficulties in its detection
the amino acid ß-alanine, however, did not find much attention (75).

Ornithine is no original component of milk; higher amounts of this substance
are formed from arginine residues only at alkaline pH where the mechanism

follows a reaction of pseudo-zero order (68). At a given pH value thus ornithine
can be used as parameter for the heat treatment of proteins.

Finally the formation of so-called isopeptide linkages - amide linkages
between the side chains of aspartic acid and glutamic acid on the one hand
and lysine residues on the other hand - belong to the undesirable crosslinking
reactions because they reduce the protein digestibility (76). Isopeptides
are eluted within an ion exchange chromatogram at characteristic positions;
the analytical problem is the preparation of the sample, namely its enzymatic
hydrolysis (77, 78). Since isopeptides are only formed during strong heating
of proteins in mediums of low humidity they are no serious problem in milkproducts - although there are contradictory opinions on this subject (77, 79).

Without doubt the question of enzymatic total or partial hydrolysis of proteins
for analytical work is still a very interesting field for research - although
a high level of knowledge has already been reached (78). In this connexion
one should think of the analysis of glycoproteins and their changes during
processing of milk and even more of the characterization of the Maillard
reaction.

Our knowledge of this reaction is mainly based on model reactions. It is
time to investigate the postulated stages of the reaction with proteins.
At present only furosine and pyridosine as reaction products of carbohydrate
with the amino acid lysine can be analysed routinely - both are merely artefacts
from the acid hydrolysis of Maillard-damaged proteins (80). In this connexion
the still unsolved problem of the formation of humin during the hydrolysis
of samples rich in carbohydrates should be mentioned leading to negative
consequences for the analysis.

This phenomenon interferes especially in the determination of the so-called
available lysine. There are numerous methods partly based on the substitution
of the  -amino group with marker groups, partly based on the dye binding
capacity of the non-substituted groups (70). All these methods have the
disadvantage of a bad comparability - particularly with biological tests
- and of inter-laboratory variations. Some problems of the substitution
technique can be eliminated by using HPLC instead of simple photometry (81).
Another progress would be if the matrix of non-protein material could simply
be decomposed enzymatically and removed by dialysis (82).

In the evaluation of food protein by means of chemical methods reactive
lysine has got the function of an indicator substance; partly this function
results from the fact that already in former times rather simple although
in some cases not satisfactory procedures for the determination of lysine
were available.Because of this exceptional position of lysine it would be
desirable if at least for the other essential amino acids - e. g. methionine
and tryptophan - simple, rapid and reliable methods were available. Chemical
digestibility tests that correlate with the biological results better than
the present tests do would also be helpful for the evaluation of technological
processes used in the dairy industry.

New techniques of analysis - e. g. HPLC and computer-controlled mass spectrometry of peptides - and transfer of ways of thinking from the wide field of
the analysis of posttranslational changes in proteins (83) will not only
help to identifiy still unknown artefacts in milk proteins; they will especially
help to associate phenomena   observed to certain positions in the molecules
and to show interrelations between chemical structure and artefact, i. e.
to make the undesired reactions more comprehensible and hence more controllable.

That is the reason for our research on artefacts in milk and milkproducts.
This paper could only incompletely indicate the aims and could only partly
mention the ways to reach these aims. If it has aroused curiosity of discovering
a not recognizable aim and of going new ways it has fulfilled its purpose.

References .

1. WIECHEN, A. & KNOOP, A.-M. (1978). Milchwiss. 33, 213

2. NAKAZAWA, Y., OISHI, H., UCHIDA, T. & OZAKI, M. (1982). Jap. J. Dairy
   Food Sci. 31, A-113

3. DIERMAIER, P. & KLOSTERMEYER, H., unpublished

4. KNOOP, A.-M., FREDE, E. & PRECHT, D. (1978). Milchwiss. 33, 696

5. FRITSCH, R. J. (1982). Thesis, TU München

6. FRITSCH, R. J., HOFFMANN, H. & KLOSTERMEYER, H. (1980). Z. Lebensm. Unters.
   Forsch. i76, 341

7. KLOSTERMEYER, H. & GEIER, H. (1983). Deutsche Milchwirtschaft 34, 1667

8. DE RHAM, O. (1983). Personal Communication

9. KIRCHMEIER, O. (1976). Z. Lebensm. Unters. Forsch. 160, 293

10. GROSCH, W. (1984). Lebensmittelchem. Gerichtl. Chem. 38, in press

11. O'CONNOR, D. E., MIHELICH, E. D. & COLEMAN, M. C. (1981). J. Am. Chem. Soc. 103, 223

12. TIMMEN, H. & VOSS, E. (1975). Milchwiss. 30, 199

13. INT. DAIRY FEDERATION (1975). Proceedings of the lipolysis symposium, Cork, IDF-Doc Nr. 86

14. SHUKLA, V. K. S., SCHIØTZ-NIELSEN, W. & BATSBERG, W. (1983). Fette, Seifen, Anstrichm. 85, 274

15. PATTON, G. M., FASULO, J. M. & SANDER, J. R. (1982). J. Lipid Res. 23, 190

16. BACK, W.-D. & REUTER, H. (1973). Milchwiss. 28, 284

17. FINK, A. (1984). Thesis, TU München

18. KORN, M. (1983). Deutsche Lebensm. Rundschau 79, 1

19. FREDE, E., PRECHT, D. & PETERS, K.-H. (1982). Milchwiss. 37, 657 and 733

20. MARTINEZ-CASTRO, I. & OLANO, A. (1980). Milchwiss. 35, 5

21. GEIER, H. & KLOSTERMEYER, H. (1983). Milchwiss. 38, 475

22. GEIER, H. & KLOSTERMEYER, H. (1980). Z. Lebensm. Unters. Forsch. 171, 443

23. BURTON, H. (1983). Annual Report of the Ntl. Inst. for Res. Dairy. Reading

24. KUNDER, I. Thesis, TU München, in preparation

25. BRUNN, W. & KLOSTERMEYER, H. (1983). Z. Lebensm. Unters. Forsch. 176, 108

26. BRUNN, W. & KLOSTERMEYER, H. (1983). Z. Lebensm. Unters. Forsch. 176, 367

27. LECHNER, E. (1982). Deutsche Milchwirtschaft 33, 489

28. FINK, R.(1984). Thesis, TU München

29. LECHNER, E., KUNDER, I. & KLOSTERMEYER, H., unpublished

30. DE KONING, P. J., EISSES, J. & DE VRIES, H. (1966). Neth. Milk & Dairy J. 20, 204

31. FABRIS, A. (1967). Alimentazione Animale 11, 469

32. OLIEMAN, C. & VAN DEN BEDEM, J. W. (1983). Neth. Milk & Dairy J. 37, 27

33. GOTTSCHALK, A. (1972). Glycoproteins (GOTTSCHALK, A. Ed.), Elsevier Publ. Comp., p. 141

34. HINDLE, E. J. & WHEELOCK, J. V. (1970). J. Dairy Res. 37, 389 and 397

35. LAHAV, E., EDELSTEN, D., SODE-MOGENSEN, M. T. & SOFER, E. (1971). Milchwiss. 26, 489

36. BÖNISCH, U. (1982). Thesis, TU München

37. BEUTLER, H.-O. & BEINSTINGL, G. (1982). Deutsch. Lebensm. Rundschau 78, 9

38. KLOSTERMEYER, H., RAUSCH, Ch. & LECHNER, E., unpublished

39. WOLFSCHOON-POMBO, A., KLOSTERMEYER, H. & WEISS, G. (1982). Milchwiss. 37, 80

40. WOLFSCHOON-POMBO, A. & KLOSTERMEYER, H. (1981). Milchwiss. 36, 598

41. SHALABI, S. I. & FOX, P. F. (1982). J. Dairy Res. 49, 197

42. SWEETSUR, A. W. M. & MUIR, D. D. (1981). J. Dairy Res. 48, 163

43. MCKENZIE, H.A. (1971). Milk Proteins (MCKENZIE, H. A. Ed.) Academic Press, Vol. II, p. 316

44. ITOH, T., WADA, Y. & NAKANISHI, T. (1976). Agr. Biol. Chem. 40, 1083

45. DE WIT, J. N., KLARENBEEK, G. & HONTELEZ-BACKX, E. (1983). Neth. Milk & Dairy J. 37, 37

46. WILLNER, I. (1984). Thesis, TU München

47. BUCHHEIM, W. (1982). Food. Microstructure 1, 189

48. STEININGER, L. (1984). Thesis, TU München

49. RICHARDSON, B. C. & PEARCE, K. N. (1981). New Zeeland J. Dairy Sci. Technol. 16, 209

50. ROLLEMA, H. S., VISSER, S. & Poll, J. K. (1983). Milchwiss. 38, 214

51. KROLL, S., GLEISSNER, R.~& KLOSTERMEYER, H. (1983). Milchwiss. 38, 718

52. WESTERMEIR, TH. (1984). Thesis, TU München

53. CLIFFE, A. J. & LAW, B. A. (1982). J. Dairy Res. 49, 209

54. KROLL, S. & KLOSTERMEYER, H. (1984). Z. Lebensm. Unters. Forsch., in press

55. REIMERDES, E. H. & HERLITZ, E. (1979). J. Dairy Res. 46, 219

56. ANDREWS, A. T. & ALICHANIDIS, E. (1983). J. Dairy Res. 50, 275

57. MROWETZ, G. & KLOSTERMEYER, H. (1972). Z. Lebensm. Unters. Forsch. 149, 74 and 133

58. LECHNER, E. & KLOSTERMEYER, H. (1981). Milchwiss. 36, 267

59. DE KONING, P. J. & VAN ROOIJEN, P. J. (1971).    Milchwiss. 26, 1

60. DE KONING, P. J., VAN ROOIJEN, P. J. & DRAAISMA, J. TH. M. (1976). Milchwiss. 31, 261

61. WIECHEN, A. & KNOOP, A.-M. (1974). Milchwiss. 29, 65

62. SCHMIDT, D. G. & BUCHHEIM, W. (1980). Proc. 7th Europ. Congr. Electron Microscopy 2, 620

63. GUTHY, K., HONG, Y.-H. & KLOSTERMEYER, H. (1983). Milchwiss. 38, 321

64. SCHNACK, U. & KLOSTERMEYER, H. (1983). Milchwiss. 35, 206

65. WATANABE, K. & KLOSTERMEYER, H. (1976). J. Dairy Res. 43, 411

66. MANSON, W. & CAROLAN, TH. (1972). J. Dairy Res. 39, 189

67. WATANABE, K. & KLOSTERMEYER, H. (1977). Z. Lebensm. Unters. Forsch. 164, 77

68. MEYER, M. (1982). Thesis, TU München.

69. FRIEDMAN, M. & MASTERS, P. M. (1982). J. Food Sci. 47, 760

70. BODWELL, C. E., ADKIN, J. S. & HOPKINS, D. T. (Eds.) (1981). Protein Quality in Humans, Avi Publ. Corp. Inc.

71. KLOSTERMEYER, H. & REIMERDES, E. H. (1977). Advan. Exp. Med. Biol. 86B, 263

72. FÖHLES, J., SOUREN, I. & KLOSTERMEYER, H. (1976). Fed. Lainiere Int., Paris, Report Nr. 8

73. DE KONING, P. J. & VAN ROOIJEN, P. J. (1982) J. Dairy Res. 49, 725

74. FRITSCH, R. J. & KLOSTERMEYER, H. (1981). Z. Lebensm. Unters. Forsch. 172, 435

75. FRITSCH, R. J. & KLOSTERMEYER, H. (1981). Z. Lebensm. Unters. Forsch. 173, 101

76. ASQUITH, R. S., OTTERBURN, M. S. & SINCLAIR, W. J. (1974). Angew. Chem. 86, 580

77. SCHMITZ, I., ZAHN, H., KLOSTERMEYER, H., RABBEL, K. & WATANABE, K. (1976). Z. Lebensm. Unters. Forsch. 160, 377

78. RÖPER, K., FÖHLES, J. & KLOSTERMEYER, H. (1984). Methods Enzymol. 107, in press

79. HURREL, R. F., CARPENTER, K. J., SINCLAIR, W. J., OTTERBURN, M. S. & ASQUITH, R. S. (1976). Br. J. Nutr. 35, 383

80. STEINIG, J. & MONTAG, A. (1982). Z. Lebensm. Unters. Forsch. 174, 453

81. WUNDENBERG, K. & KLOSTERMEYER, H. (1983). Kieler Milchwirtsch. Forschungsber. 35, 315

82. WUNDENBERG, K., Thesis, in preparation

83. COLOWICK, S. P. & KAPLAN, N. O. (Eds.) (1984). Methods Enzymol. 107, in press

# Bioluminescence Assay to Detect Antibiotics and Antiseptics in Milk

By R. Quesneau*, M. Bigret, and F. M. Luquet
S.S.H.A., RUE DU CHEMIN BLANC, B.P. 138, CHAMPLAN, 91160 LONGJUMEAU,
FRANCE

INTRODUCTION

The French official method for the research of inhibitors
in milk (including antibiotics and sulfonamids) consists
in two steps ; control of acidification  and control by
diffusion (Galesloot) (1).

For the first step one incubates Streptococcus thermophilus TJ
within the sample in presence of yeast extract and bromocresol
purple as color indicator during two and half hours.

This technique is simple and sensitive but it gives information
about the result of a metabolic activity without precision about
the real nature of the metabolites produced.

The use of the ATP assay to measure a variety of microbiocidal
agents and antibiotics has been reported (2, 3, 4, 5, 6, 7)
including in milk (8).

It has the interest to be a direct control of metabolic level
of test microorganisms. The purpose of our work was to compare
the classical measurements of S. thermophilus acidification
(Dornic degree, coagulation ...) with this new technique.

MATERIALS AND METHODS :

Test culture

Streptococcus thermophilus TJ was obtained from CNRZ Laboratory,
Jouy en Josas, France.
The procedure followed was the method of the CNERNA (Centre
National de Coordination des Etudes et Recherches sur la
Nutrition) as referenced in CNERNA instruction of inhibitor
in milk. (9).

Chemicals

Raw milk was supplied directly by farmers. 120 g of milk powder
(commercially available) were added to 930 g of tap water to
obtain reconstituted milk. Antibiotics and sulfonamids were
purchased from SIGMA. Chemicals for culture medium as yeast
extract were obtained from DIFCO.

Extraction procedure

The method used was dimethyl sulfoxide extraction previously
described by HYSERT et al.(10) and studied by JAKUBCZAK et
al. (11). 0.1 ml of the sample was pipetted with semi automatic
pipette into a clean glass tube filled with 0.9 ml of pure DMSO.
The tube was shaken on a Vortex mixer during 1 min. 4 millili-
ters of TRIS buffer was added, and the extracted sample was
mixed 15 sec. before being titrated.

Luciferase assay

ATP was determined by means of the Luciferase reaction (12) with
a BIOCOUNTER M 2010 multijet®, LUMAC/3M commercially available
from SEMPA-CHIMIE, Paris, France with respect of LUMAC protocole
for bioluminescence measurement.

Inoculation culture

A reserve culture of S. Thermophilus     is incubated over-
night at 37°C till to coagulation without excess of serum.
10 ml of reconstituted milk free of inhibitor containing 20

mg/ml of yeast extract are added with 2 ml of coagulated milk
and incubated at 45°C. This culture is called inoculation culture. Its development is monitored each half-hour by biolumines-
cent assay.

Sampling was made after 60 minutes if the ATP level at this
time was increased 5 fold and bioluminescence reached
at least 10,000 to 15,000 RLU (Relative Light Units).

ATP pool variation measurement

Milks free from inhibitor were used as reference to describe the
variation of ATP pool in presence of antibiotics or sulfonamids.
These milks were called "reference milks" and added with inhibitor to prepare assays.

In each experiment the milk was inoculated with the same quantity of bacteria after control of inoculation culture.

ATP pool at different times was compared to ATP pool 10 min after
inoculation.

Calculation

ATP pool variation of assays was compared to that of corresponding "reference milk".

For this purpose the ratio ATP pool at t time/ATP pool at 10
minutes (expressed in RLU) was defined as Pt. And the ratio Pt
of the assay/Pt of the reference was defined as Rt. A statistical
study shows that the values' variation around the average is 10 %
for the reference, so less than a 20 % decrease of the assay
compared with reference (Rt between 1 and 0.8) is not considered
as inhibition.
Consequently values of Rt $\leq$ 0.8 indicated a significant
difference between the assay and the reference.

Knowing that the measurement of acidity by the evaluation of the
Dornic degree is commonly use as criteria of fermentescibility,
the assays were measured also by this means and compared by the
use of identical ratios.

Influence of antibiotics

The following antibiotics were added to reconstituted milk or to
raw milk at different concentrations :
Penicillin G, dihydrostreptomycin, tetracyclin, chloramphenicol
and erythromycin. Sulfadiazine was also assayed in milk containing 10 $\mu$g/ml of trimethoprim.

The corresponding figures show the variations of ATP pool and
Dornic degree of cultures in presence of these inhibitors compared with those of a reference milk expressed as Pt = f(t).


RESULTS AND DISCUSSION

Several antibiotics act by lysis of microorganisms; cytoplasm
content could diffuse out of cells in these cases. A diffusion
of ATP in culture medium was observed in presence of Penicillin
by simple dosage of free ATP by bioluminescence, without
extraction of bacteria.

With the aim of measuring only the intracellular ATP content of
S. Thermophilus an experiment was performed.

It consisted of adding apyrase in milk used for culture and
comparing the influence of this addition on the sensibility of the
test.

The ATP content of a bacterial extract was measured in presence
of 2 units of apyrase per ml of reconstituted milk.

The results obtained show that the enzyme did not modify the
evolution of the strain in milk free of inhibitor  and that
free ATP was destroyed in two minutes of incubation at 25°C.

After that the stability of the extracts made with DMSO was
demonstrated as well as the inhibition of previously added
apyrase.
This phenomenon was soon noted by SHAW (13)  with other
enzymes.

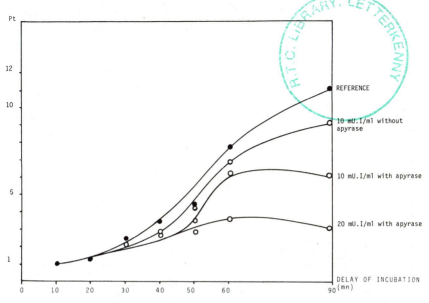

FIGURE 1    ATP pool variation of S. Thermophilus TJ culture in reconstituted milk containing Penicillin G.

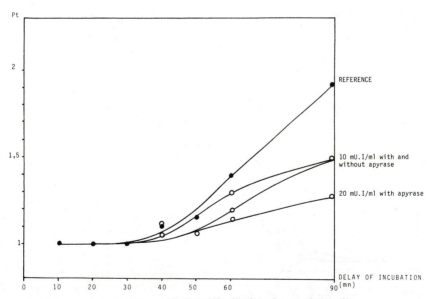

FIGURE 2    Dornic degree variation of S. Thermophilus TJ culture in reconstituted milk containing Penecillin G.

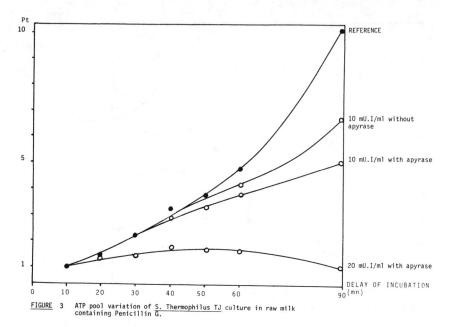

FIGURE  3    ATP pool variation of S. Thermophilus TJ culture in raw milk
             containing Penicillin G.

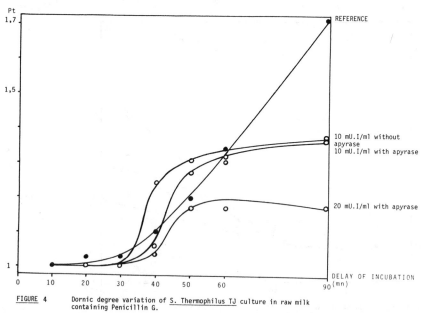

FIGURE  4    Dornic degree variation of S. Thermophilus TJ culture in raw milk
             containing Penicillin G.

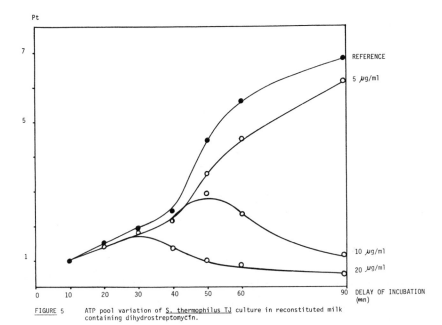

FIGURE 5     ATP pool variation of <u>S. thermophilus TJ</u> culture in reconstituted milk containing dihydrostreptomycin.

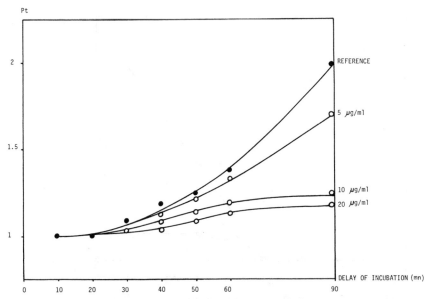

FIGURE 6     Dornic degree variation of <u>S. thermophilus TJ</u> culture in reconstituted milk containing dihydrostreptomycin.

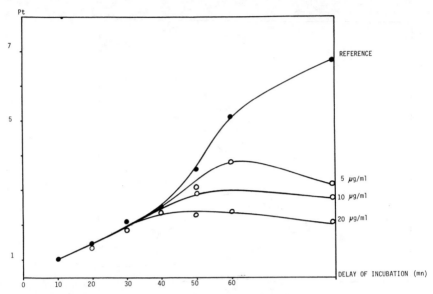

<u>FIGURE</u> 7    ATP pool variation of <u>S. Thermophilus TJ</u>  culture in raw milk containing
dihydrostreptomycin.

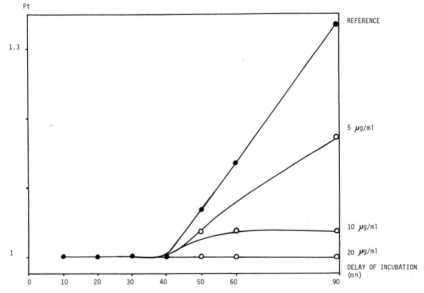

<u>FIGURE</u> 8    Dornic degree variation of S. <u>Thermophilus TJ</u> culture in raw milk containing
dihydrostreptomycin.

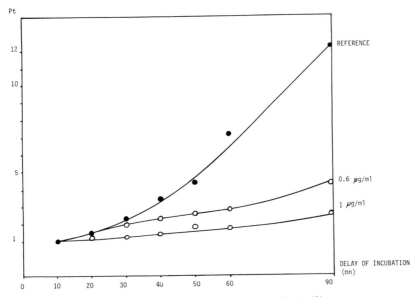

<u>FIGURE</u> 9    ATP pool variation of <u>S. Thermophilus TJ</u> culture in reconstituted milk
containing tetracycline.

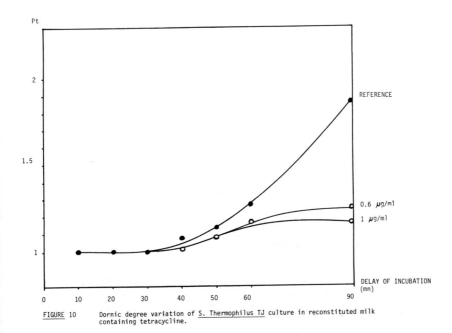

<u>FIGURE</u> 10    Dornic degree variation of <u>S. Thermophilus TJ</u> culture in reconstituted milk
containing tetracycline.

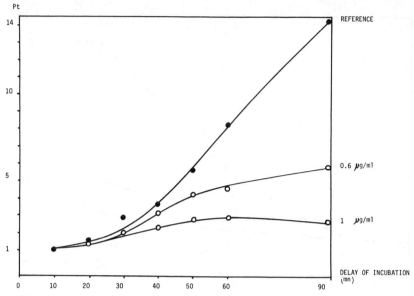

<u>FIGURE</u> 11    ATP pool variation of <u>S. Thermophilus TJ</u> culture in raw milk containing tetracycline.

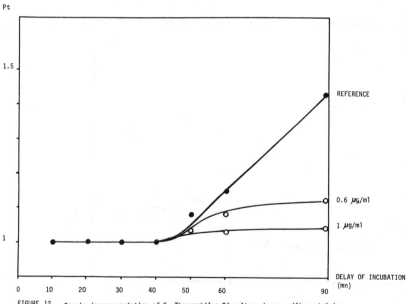

<u>FIGURE</u> 12    Dornic degree variation of <u>S. Thermophilus TJ</u> culture in raw milk containing tetracycline.

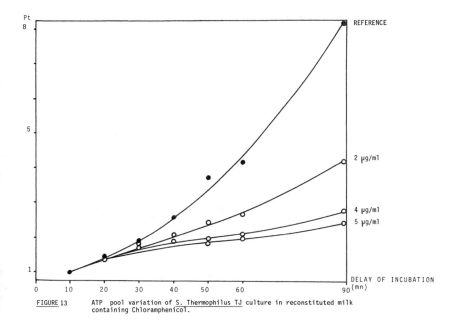

FIGURE 13    ATP   pool variation of S. Thermophilus TJ culture in reconstituted milk
containing Chloramphenicol.

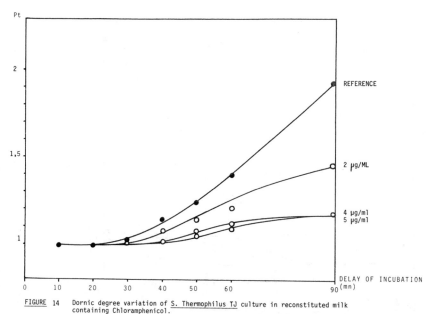

FIGURE 14    Dornic degree variation of S. Thermophilus TJ culture in reconstituted milk
containing Chloramphenicol.

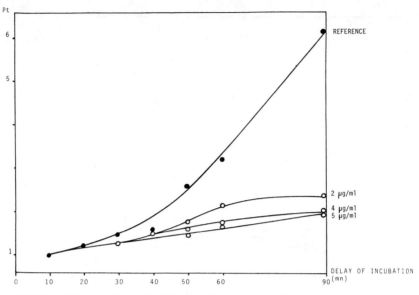

ATP pool variation of S. Thermophilus TJ culture in raw milk
containing Chloramphenicol.

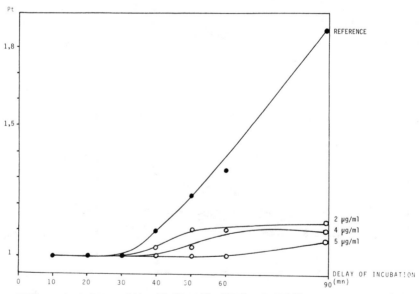

FIGURE 16     Dornic degree variation of S. Thermophilus TJ culture in raw milk
containing Chloramphenicol.

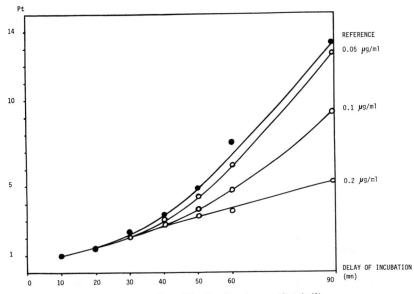

FIGURE 17   ATP pool variation of S. Thermophilus TJ culture in reconstituted milk
containing erythromycin.

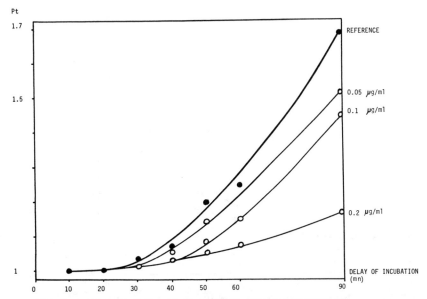

FIGURE 18   Dornic degree variation of S. Thermophilus TJ culture in reconstituted milk
containing erythromycin.

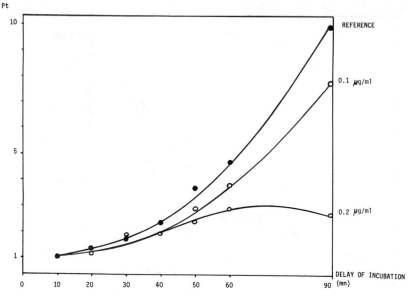

FIGURE 19    ATP pool variation of <u>S. Thermophilus TJ</u> culture in raw milk containing erythromycin.

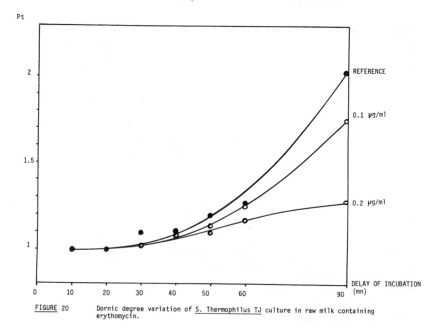

FIGURE 20        Dornic degree variation of <u>S. Thermophilus TJ</u> culture in raw milk containing erythomycin.

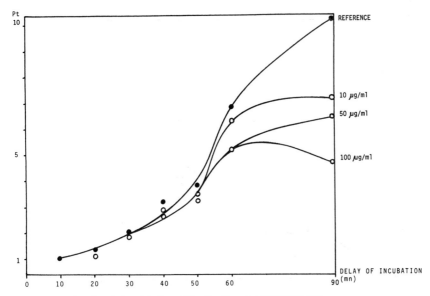

FIGURE 21  ATP pool variation of S. Thermophilus TJ culture in reconstituted milk containing Sulfadiazine (trimethoprim)

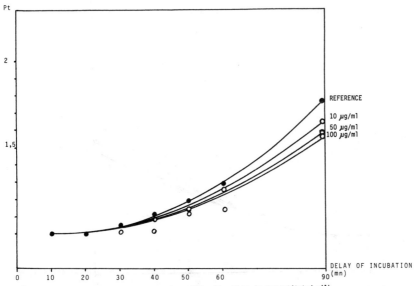

FIGURE 22  Dornic degree variation of S. Thermophilus TJ culture in reconstituted milk containing Sulfadiazine (trimethoprim).

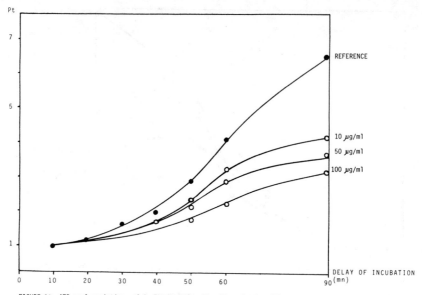

FIGURE 23   ATP pool variation  of S. Thermophilus TJ culture in raw milk
            containing Sulfadiazine (trimethoprim).

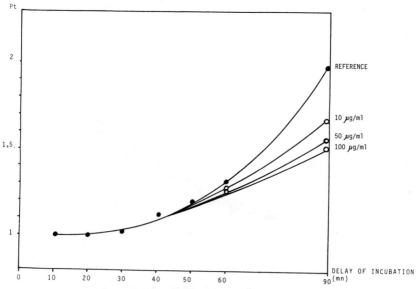

FIGURE 24   Dornic degree variation of S. Thermophilus TJ culture in raw milk
            containing Sulfadiazine (trimethoprim)

Figure 3 shows the results otained. It was clear that
within 40 minutes it is possible to detect 10 mU/ml of
Penicillin with apyrase whereas without enzyme this is possible
only within 90 minutes.

The acidity development of milk during S.Thermophilus incubation
was never important.

It started from about 35°D in reconstituted milk and 25°D in
raw milk and reached no more than twice its initial value within
90 minutes.

In the same time, the ATP pool reached 7 to 10 fold its initial
level. After 40 minutes of incubation the increase of these
two parameters was respectively 1.1 to 1.2 fold and 2 to 3
fold.

Despite this difference it is evident that the ATP parameter evo-
luated similarly to the development of acidity in a reference
milk or in presence of antibiotics.

This is not the case with sulfonamids. When the ATP level slowed
down the acidity continued to evoluate. The bacteriostatic effect
of these products could explain this phenomenon.

The concentrations of antibiotics used were those that prevented
the coagulation of milk within two and half hours in our experimental conditions, except for 0.05 μg/ml of erythromycin with
which the coagulation occurred after three hours.

About the results obtained, there was no significant difference
between raw milk and reconstituted milk.

Within 40 minutes the profile of ATP was flat with concentrations
equal or exceeding 20 mU/ml of penicillin, 20 μg of dihydrostreptomycin, 1 μg of tetracyclin, 2 μg of chloramphenicol or 0.4 μg
of erythromycin.

In such cases it is not necessary to calculate Rt values to be sure
of the antibiotic's presence in assays. With lower concentrations

of antibiotic an ATP control after 60 minutes always gave a result
more accurate than Dornic degree.

The Tables 1 to 12 show the results obtained  using the Rt ratio
defined above.

ATP control  informed the operator on antibiotic suspicion at
least 30 minutes before the acidity dosage and the effect of
antibiotics and sulfonamids is more evident on ATP pool than on
acid production.

During the different experiments it was demonstrated that
apyrase addition in the samples was specially efficient to
detect small quantities of penicillin and dihydrostreptomycin.
With low concentration of chloramphenicol (less than $1\mu g/ml$)
the ATP pool per cell tended to increase. The consequence was
a limited sensitivity of the test for such concentrations of
this antibiotic.

Generally it was possible to work with different milks of the
same nature (raw or reconstituted) as reference or for the
preparation of assays.

However several reference milks (selected by the official method)
gave results which could be confused after 40 minutes of incubation with those of milks containing low concentration of antibiotics. To avoid this problem it is necessary not to use only
one reference milk for a series of tests, but to do an average
with three of them and to use the Pt values obtained for the
rest of calculations.

| METHOD USED | BIOLUMINESCENCE | | | DORNIC DEGREE | | |
|---|---|---|---|---|---|---|
| Antibiotic concentration (mU.I/ml) | 10 - A | 10 + A | 20 + A | 10 - A | 10 + A | 20 + A |
| 10 | 1 | 1 | 1 | 1 | 1 | 1 |
| 20 | 0.87 | ＞ 1 | 0.93 | 1 | 1 | 1 |
| 30 | 0.84 | 0.81 | 0.81 | 1 | 1 | 1 |
| 40 | 0.85 | 0.77 | 0.75 | 0.95 | ＞ 1 | 0.95 |
| 50 | 0.97 | 0.78 | 0.64 | 0.93 | 1 | 0.95 |
| 60 | 0.89 | 0.80 | 0.47 | 0.85 | 0.92 | 0.82 |
| 90 | 0.80 | 0.54 | 0.26 | 0.78 | 0.78 | 0.66 |

The left axis of the table reads "DELAY OF INCUBATION (mn)".

TABLE 1  Rt values during incubation for different concentrations of Penicillin G added in reconstituted milk.

| METHOD USED | BIOLUMINESCENCE | | | DORNIC DEGREE | | |
|---|---|---|---|---|---|---|
| Antibiotic concentration (mU.I/ml) | 10 - A | 10 + A | 20 + A | 10 - A | 10 + A | 20 + A |
| 10 | 1 | 1 | 1 | 1 | 1 | 1 |
| 20 | ＞ 1 | 0.93 | 0.92 | 0.97 | 0.97 | 0.97 |
| 30 | ＞ 1 | ＞ 1 | 0.68 | 0.97 | 0.97 | 0.97 |
| 40 | 0.91 | 0.88 | 0.55 | 0.90 | ＞ 1 | 0.88 |
| 50 | 0.93 | 0.87 | 0.43 | ＞ 1 | ＞ 1 | 0.97 |
| 60 | 0.87 | 0.79 | 0.33 | 1 | 0.97 | 0.87 |
| 90 | 0.66 | 0.36 | 0 | 0.79 | 0.79 | 0.68 |

The left axis of the table reads "DELAY OF INCUBATION (mn)".

TABLE 2  Rt value during incubation for different concentrations of Penicillin G added in raw milk.

| METHOD USED | BIOLUMINESCENCE | | | DORNIC DEGREE | | |
|---|---|---|---|---|---|---|
| Antibiotic concentration (µg/ml) | 5 | 10 | 20 | 5 | 10 | 20 |
| DELAY OF INCUBATION (mn) 10 | 1 | 1 | 1 | 1 | 1 | 1 |
| 20 | 0.99 | 0.97 | >1 | 1 | 1 | 1 |
| 30 | 0.87 | 0.97 | 0.97 | 0.97 | 0.99 | 0.95 |
| 40 | 0.85 | 0.87 | 0.59 | 0.89 | 0.90 | 0.87 |
| 50 | 0.77 | 0.65 | 0.21 | 0.98 | 0.91 | 0.87 |
| 60 | 0.80 | 0.41 | 0.16 | 0.97 | 0.86 | 0.82 |
| 90 | 0.90 | 0.16 | 0.09 | 0.88 | 0.64 | 0.61 |

TABLE 3   Rt values during incubation for different concentrations of dihydrostreptomycin (+ apyrase) added in reconstituted milk.

| METHOD USED | BIOLUMINESCENCE | | | DORNIC DEGREE | | |
|---|---|---|---|---|---|---|
| Antibiotic concentration (µg/ml) | 5 | 10 | 20 | 5 | 10 | 20 |
| DELAY OF INCUBATION (mn) 10 | 1 | 1 | 1 | 1 | 1 | 1 |
| 20 | 0.91 | 0.78 | 1 | 1 | 1 | 1 |
| 30 | 0.89 | 0.82 | 0.92 | 1 | 1 | 1 |
| 40 | 0.91 | 0.95 | 0.96 | 1 | 1 | 1 |
| 50 | 0.84 | 0.81 | 0.62 | 0.97 | 0.97 | 0.94 |
| 60 | 0.73 | 0.48 | 0.46 | 1 | 0.91 | 0.88 |
| 90 | 0.47 | 0.42 | 0.30 | 0.87 | 0.77 | 0.75 |

TABLE 4   Rt values during incubation for different concentrations of dihydrostreptomycin (+ apyrase) added in raw milk.

| METHOD USED | BIOLUMINESCENCE | | DORNIC DEGREE | |
|---|---|---|---|---|
| Antibiotic concentration (µg/ml) | 0.6 | 1 | 0.6 | 1 |
| 10 | 1 | 1 | 1 | 1 |
| 20 | 0.99 | 0.82 | 1 | 1 |
| 30 | 0.80 | 0.69 | 1 | 1 |
| 40 | 0.64 | 0.49 | 0.93 | 0.87 |
| 50 | 0.57 | 0.40 | 0.95 | 1 |
| 60 | 0.39 | 0.28 | 0.91 | 0.91 |
| 90 | 0.35 | 0.23 | 0.67 | 0.62 |

(DELAY OF INCUBATION (mn))

TABLE 5   Rt values during incubation for different concentrations of tetracycline added in reconstituted milk.

| METHOD USED | BIOLUMINESCENCE | | DORNIC DEGREE | |
|---|---|---|---|---|
| Antibiotic concentration (µg/ml) | 0.6 | 1 | 0.6 | 1 |
| 10 | 1 | 1 | 1 | 1 |
| 20 | 0.86 | 0.86 | 1 | 1 |
| 30 | 0.71 | 0.67 | 1 | 1 |
| 40 | 0.85 | 0.43 | 1 | 1 |
| 50 | 0.74 | 0.47 | 0.95 | 0.95 |
| 60 | 0.54 | 0.33 | 0.93 | 0.89 |
| 90 | 0.39 | 0.18 | 0.78 | 0.72 |

(DELAY OF INCUBATION (mn))

TABLE 6   Rt values during incubation for different concentrations of tetracycline added in raw milk.

| METHOD USED | BIOLUMINESCENCE | | | DORNIC DEGREE | | |
|---|---|---|---|---|---|---|
| Antibiotic concentration (µg/ml) | 2 | 4 | 5 | 2 | 4 | 5 |
| 10 | 1 | 1 | 1 | 1 | 1 | 1 |
| 20 | 0.93 | 0.95 | 1 | 1 | 1 | 1 |
| 30 | 0.89 | 0.87 | 0.94 | 0.99 | 0.98 | 1 |
| 40 | 0.79 | 0.70 | 0.70 | 0.93 | 0.88 | 0.87 |
| 50 | 0.65 | 0.52 | 0.47 | 0.92 | 0.85 | 0.87 |
| 60 | 0.63 | 0.50 | 0.53 | 0.86 | 0.79 | 0.78 |
| 90 | 0.49 | 0.32 | 0.31 | 0.75 | 0.61 | 0.61 |

(DELAY OF INCUBATION (mm))

TABLE 7   Rt values during incubation for different concentrations of chloramphenicol added in reconstituted milk.

| METHOD USED | BIOLUMINESCENCE | | | DORNIC DEGREE | | |
|---|---|---|---|---|---|---|
| Antibiotic concentration (µg/ml) | 2 | 4 | 5 | 2 | 4 | 5 |
| 10 | 1 | 1 | 1 | 1 | 1 | 1 |
| 20 | 0.91 | 1 | 0.90 | 1 | 1 | 1 |
| 30 | 0.87 | 0.89 | 0.87 | 1 | 1 | 1 |
| 40 | 0.92 | 0.96 | 0.90 | 0.93 | 0.93 | 0.90 |
| 50 | 0.68 | 0.60 | 0.56 | 0.91 | 0.83 | 0.81 |
| 60 | 0.61 | 0.54 | 0.50 | 0.84 | 0.82 | 0.75 |
| 90 | 0.38 | 0.34 | 0.31 | 0.72 | 0.60 | 0.57 |

(DELAY OF INCUBATION (mm))

TABLE 8   Rt values during incubation for different concentrations of chloramphenicol added in raw milk.

| METHOD USED | BIOLUMINESCENCE | | | DORNIC DEGREE | | |
|---|---|---|---|---|---|---|
| Antibiotic concentration (μg/ml) | 0.05 | 0.1 | 0.2 | 0.05 | 0.1 | 0.2 |
| 10 | 1 | 1 | 1 | 1 | 1 | 1 |
| 20 | 1 | 1 | 1 | 1 | 1 | 1 |
| 30 | 0.98 | 0.83 | 0.85 | 1 | 1 | 1 |
| 40 | 0.95 | 0.86 | 0.74 | 1 | 1 | 1 |
| 50 | 0.93 | 0.79 | 0.69 | 0.95 | 0.92 | 0.89 |
| 60 | 0.83 | 0.65 | 0.47 | 0.93 | 0.93 | 0.86 |
| 90 | 0.96 | 0.71 | 0.40 | 0.89 | 0.86 | 0.69 |

(Rows labeled under DELAY OF INCUBATION (mn))

TABLE 9   Rt values during incubation for different concentrations of erythromycin added in reconstituted milk.

| METHOD USED | BIOLUMINESCENCE | | | DORNIC DEGREE | | |
|---|---|---|---|---|---|---|
| Antibiotic concentration (μg/ml) | 0.05 | 0.1 | 0.2 | 0.05 | 0.1 | 0.2 |
| 10 | 1 | 1 | 1 | 1 | 1 | 1 |
| 20 | 1 | 1 | 1 | 1 | 1 | 1 |
| 30 | 1 | 1 | 1 | 0.94 | 0.95 | 0.94 |
| 40 | 0.94 | 0.96 | 0.84 | 1 | 1 | 1 |
| 50 | 1 | 0.70 | 0.67 | 1 | 1 | 0.92 |
| 60 | 1 | 0.78 | 0.60 | 1 | 1 | 0.92 |
| 90 | 1 | 0.79 | 0.28 | 0.90 | 0.86 | 0.63 |

(Rows labeled under DELAY OF INCUBATION (mn))

TABLE 10   Rt values during incubation for different concentrations of erythromycin added in raw milk.

| METHOD USED | BIOLUMINESCENCE | | | DORNIC DEGREE | | |
|---|---|---|---|---|---|---|
| Antibiotic concentration (µg/ml) | 10 | 50 | 100 | 10 | 50 | 100 |
| 10 | 1 | 1 | 1 | 1 | 1 | 1 |
| 20 | > 1 | 0.84 | 0.85 | 1 | 1 | 1 |
| 30 | 0.91 | 0.82 | 0.92 | 1 | 0.98 | 1 |
| 40 | 0.86 | 0.94 | 0.81 | 1 | 1 | 0.94 |
| 50 | 0.89 | 0.88 | 0.88 | 0.94 | 0.93 | 0.95 |
| 60 | 0.91 | 0.73 | 0.75 | 1 | 0.86 | 0.96 |
| 90 | 0.67 | 0.62 | 0.44 | 0.93 | 0.89 | 0.88 |

(left axis label: DELAY OF INCUBATION (mn))

TABLE 11  Rt values during incubation for different concentrations of sulfadiazine + trimethoprim added in reconstituted milk.

| METHOD USED | BIOLUMINESCENCE | | | DORNIC DEGREE | | |
|---|---|---|---|---|---|---|
| Antibiotic concentration (µg/ml) | 10 | 50 | 100 | 10 | 50 | 100 |
| 10 | 1 | 1 | 1 | 1 | 1 | 1 |
| 20 | 0.97 | > 1 | > 1 | 1 | 1 | 1 |
| 30 | > 1 | > 1 | 0.98 | 0.97 | 0.97 | 0.97 |
| 40 | 0.97 | > 1 | 0.86 | 1 | 1 | 1 |
| 50 | 0.78 | 0.71 | 0.59 | 1 | 1 | 0.94 |
| 60 | 0.80 | 0.71 | 0.57 | 0.96 | 0.94 | 0.88 |
| 90 | 0.63 | 0.57 | 0.49 | 0.86 | 0.81 | 0.79 |

(left axis label: DELAY OF INCUBATION (mn))

TABLE 12     Rt values during incubation for different concentrations of sulfadiazine + trimethoprim added in raw milk.

CONCLUSION

The ATP control of lactic strains cultures as S. Thermophilus gives similar information to acidity measurement and it has the advantage of speed and clearness.

The results obtained are more objective (no personal appreciation of color indicator change). On a technical point of view ATP control, which needs only 100 µl of milk for a test, allows work over a long period with a limited volume of sample whereas Dornic degree control needs 10 ml of product for each measure.

Another interest of this technique is that it is possible to obtain information on bacteriostatic effects of milk contaminants.

With addition of apyrase in the test the lysis of cells could be detected very quickly. We could speculate on the usefulness of this test as a tool for bacteriophage detection. As it is also possible to control strains other than the lactic ones there are other ways of development for use of ATP assay as inhibitor detection test in food industries.

1. Journal officiel de la République Française, 6/10/83 - P. 9089-9093.

2. C.I. Weber.
   EPA Bioassay Techniques and Environmental Chemistry, 1973, 119-138.

3. P.L. Brezonik, F.X. Browne, and J.L. Fox
   Water res., 1975, 9 : 155-162.

4. S. Ansehn, L. Nilson, H. Hojer, and A. Thore
   2nd Bi-Annual ATP Methodology Symposium, San Diego 1977, 189-203.

5. Y.S. Lee, and K.G. Crispen
   2nd Bi -Annual ATP Methodology Symposium, San Diego, 1977, 219-236.

6. M.J. Harber, and A.W. Asscher,
   J. Antimicrob. Chemother.
   1977, 3 : 35-41.

7. E.S. Littman
   ATP Methodology Seminar, San Diego,
   1975, 170-188.

8. D.C. Westhoff, T. Engler,
   J. Milk Food Technol,
   1975, 38(9) , 537-539.

9  C.N.E.R.N.A.
   Rev. Lait. Française
   1981, 395, 7-17.

10  D.W. Hysert, F. Kovecjes, N.M. Morrison
    J. Amer. Soc. Brew. Chem.
    1976, 34(4), 145-150.

11  C. Jakubczak, H. Leclerc
    Ann. Biol. Clin.
    1980, 38, 297-304.

12  B.L. Strehler - 1968
    Bioluminescent assay : principles and practice P. 99-181
    in D. Glick (ed).
    Methods of biochemical analysis, vol. 16.
    John Wiley and Sons. Inc. N.Y.

13  E. Shaw in Boyer P.D.
    The enzymes, $3^{rd}$ eds. , vol 1, p. 131
    Academic Press 1970

# Lipolysis in Milk and Factors Responsible for Its Development. Determination of Free Fatty Acids

By G. Bråthen
DAIRY LABORATORY, BOKS 9066, OSLO I, NORWAY

## Introduction

Lipolysis (enzymatic hydrolysis of milk fat) is one of the most important factors responsible for the flavor impairment of milk and milk products. This problem has increased, rather than decreased, since the introduction of modern methods of milking, storage, transport and processing. To unravel the causes and to find practical solutions the dairy industry needs analytical data to assess the magnitude and development of lipolysis. This constitutes a challenge to contemporary dairy analytical techniques.

In this paper I will briefly discuss factors responsible for the development of lipolysis in milk, and some considerations about sensory evaluation in this context. This will form a background for a short assessment of analytical methods now available. Some methods I find useful will be described. I hope this can form a starting point for a discussion on this important topic.

## Lipolysis of milk and factors responsible for its development

Lipolysis can be defined as the enzymatic hydrolysis of milk fat triglycerids to mono-, di- glycerides, glycerol and free fatty acids (FFA). Considerable research has been done to unravel the mechanism of and factors influencing the lipolysis of milk fat. Comprehensive reviews have been published[1,2,3,4]. It is now considered that one enzyme is mainly responsible[5]: lipoprotein lipase (E.C. 3.11.34). One important fact is that milk always contains more than enough mLPL to cause significant lipolysis[6]. As most milks do not develop severe lipolysis one or several factors must impede the action of mLPL. As a matter of fact the action of mLPL (in milk) is governed by a number of complex and often inter-

related factors that can lead to "activation" or "inhibition". It
has been common to distinguish between lipolysis which arises as a
result of mechanical damage to the fat globule ("induced lipolysis")
and so called "spontaneous lipolysis" which occurs in "undamaged"
milk that has only undergone thermal treatment. This distinction
is somewhat arbitary because of an interaction between these
mechanisms and a possibility that they occur simultaneously[7].
In this brief survey I will only mention some of the factors which
have been shown to have an effect on the development of lipolysis.
Mechanical agitation of warm milk in pipelines is likely to enhance
lipolytic activity. Foaming (air admission to pipeline milking
installations) has a greater effect on the development of lipolysis
than mere agitation[3]. Generally the effect of mechanical agitation
on lipolysis depends on the temperature history of the milk, the
type of treatment and the characteristics of milk[3]. Mechanical/
thermal treatment can also lead to a transfer of mLPL to the cream
fraction[8]. It is assumed that the intact milk fat globule membrane
(MFGM) suppresses the lipolytic activity and that the treatments
referred to above can result in damage to it (MFGM).

The addition of blood serum to milk greatly enhances lipolysis
[6,9,10]. Apparently blood serum contains activating factors that
can reach the milk through leakage. Indeed, blood serum high
density lipoproteins (HDL) have been determined immunologically
in normal milk[11], and they have been shown to activate mLPL[12].
More specifically an apo-C HDL has been shown[13,14] to activate
mLPL *in vitro*. In normal milk, however, lipolysis does not depend
solely on the amount of lipase activator[7]. A prerequisite for
activation by apo-C HDL is a physicochemical alteration of the fat
globule (and a redistribution of mLPL)[13]. This alteration can be
caused by mechanical damage, serum lipid components or heparin.
It is tempting to speculate that activator from serum apo-C HDL,
lipids and heparin mediate (at least partly) their effect through
making substrate available for mLPL. Factors known to enhance
lipolysis such as stage of lactation,[15] mastitis,[16,17], oestrus[9,10,18], diet[10,19,20,21] might primarily influence how much lipase
activator (or additional enzymes) reaches the milk. Stability
of the fat globule may, however, be directly influenced by diet[22].
This could in turn have an effect on the milks susceptibility to
lipolysis.

When lipolysis is induced by mechanical agitation (i.e. homogenization) a rapid rise of FFA-content followed by a leveling-off occurs[23]. A re-initiation of lipolysis follows after repeated homogenization[23]. The most likely explanation is that inhibitory FFA accumulates at the exposed fat globule interface (substrate inhibition)[24]. It was early suspected[25] that milk contains a factor that inhibits lipolysis. A few years ago it was demonstrated[26] that proteose peptone (especially component 3) is a strong inhibitor of lipolysis. It was concluded that comparatively small increases in proteose peptone could have a significant effect on lipolysis. The most probable mechanism is that proteose peptone acts by protecting the substrate (exposed milk fat) against mLPL attack. From the research briefly referred to above there seems to crystalize that the concept "substrate availability" is very central. The factors governing the availability of fat (triglycerides) seem to be more important than the activators/inhibitors of the m-lipoprotein-lipase per se.

One thing about lipase action that is quite important both from a sensorial and analytical point of view is the specificity of the enzyme, that is which fatty acids are released from triglycerides during lipolysis. It has been found[27] that lipo-protein lipase almost only hydrolyzes the primary ester groups of triglycerides (this can be due to the physical property of the substrate rather than a fatty acid specificity[28]). The short chain fatty acids are mainly esterified in these positions while the others are partially randomly distributed[29]. During lipolysis a relatively larger amount of lower fatty acids will be released compared with the fatty acid composition of milk fat. Is is also possible that a relatively larger proportion of short fatty acids is released during the earlier phases of lipolytic attack than later. The short fatty acids are most important sensorially while usual analytic methods extract these incompletely. These are facts that it is important to consider when interpreting and comparing analytical and sensorial data.

## Considerations about sensory evaluation

The sensory defect (rancidity) that develops during lipolysis in milk is mainly caused by lower fatty acids ($C_4$-$C_{12}$) released into the water phase. The other free fatty acids ($C_{14}$-$C_{20}$) are the major acids detected in the usual analytical methods. This

simple fact is one of the important reasons for the impossibility
of a complete correlation between acid degree values (ADV) and
flavor.  Another problem in this context is to define exactly what
a purely lipolyzed flavor is.  This depends on the level of lipolysis, other possible taints, the sensitivity of the human taste
organs in question *etc*.  Studies[30] with highly trained experts
provided with reference  samples have given correlation coefficients *versus* BDI-values of 0.67-0.82, *versus* copper soap method-
values of 0.82-0.83 (sample FFA ranged from *ca*. 0.3 to 2 times
the observed taste threshold). Other studies have tried to define
a taste threshold; here again higly trained panels were tested
with the following results: *ca*. 1.6 meq/100 g fat[30] and *ca*. 1.9
meq/100 g fat[31].  (Both referring to BDI results).  We have had
examples of non-trained panels with taste-thresholds at about 3.0
meq/100 g fat; graders at dairies usually have thresholds at about
2.2 meq/100 g fat.

My conclusion is that maybe a too high emphasis has been placed
on a very high correlation between results from analytical methods
(FFA conc.) and sensory score.  Even if a good correlation with
sensory assessment is clearly an asset for an analytical method
one must bear in mind that two different qualities are measured.
Sensorially the sensation is of rancid flavor while the analytical
method should measure the concentration of lipolysis products
(FFA).  For an analytical method information about the specificity
and recovery of FFA is most important. ( Sensory assessment can be
considered an analytical method that gives good information, but
with in most cases an undefined specificity and recovery).

Choice of analytical method
What do we want from an analytical method for determining the
extent of lipolysis, more specifically FFA? It should determine
all of the FFA in the sample, be accurate and reproducible, not
be interfered by other substances, *i.e*. lactic acid, be simple to
perform, have a high throughput of samples *etc*. Obviously not
one single analytical method can fullfill all these requirements.
Rather one must define what is needed for a particular problem and
thereby the most important attributes of the method.  In the following I will concentrate on three general methods that should be
useful for investigations of many aspects of lipolysis:

1. A simple and rapid routine method that can be performed at comparatively poorly equipped laboratories. High sample throughput is not necessary, but it should be reproducible and reliable. It should be possible to perform analyses at the place where the samples are drawn.
2. An automated method for central laboratories with high sample throughputs. Precise information about specificity, recovery and possible interferences must be established. It should be coupled with good sample transport procedures so samples can be collected and sent without significant lipolysis during the time between sampling and analysis.
3. A reference/research method that will determine not only total FFA, but the concentration of the different FFA. Recovery must be reproducible and known and interferences eliminated.

### Analytical methods for the determination of FFA

Several methods have been proposed[32] for the determination of FFA in milk. Many of those have been rendered obsolete by the development of equipment and a better understanding of the factors involved. At present we have methods that can fit in with the requirements in the above three points.

### A simple routine method.

The most widely used routine method to present is the so called BDI-metod[33]. Several modifications of the original method have been proposed[34,35,36], both to improve repeatability and simplify the analysis. The method has been very valuable in the study of lipolysis and will probably still be in use for some time to come. I find, however, that it has some shortcomings. Firstly, the isolation of the fat with detergent and heat leads to a low recovery of FFA (see Table 1). Secondly the isolation includes warming and sometimes centrifugation (and weighing): all tedious steps and not very attractive in a routine method.

Many methods based on solvent extraction of FFA and subsequent titration have been reported[37,38,39,40,41,42]. The inherent difficulty in all these methods is to get a good recovery of FFA (especially the lower hydrophilic FFA) without extracting interferences as lactates, citrates and phospolipids.

Table 1. Percentage recovery of individual fatty acids by different extraction/titration methods.

| Fatty acid | BDI – method | | | Solvent extraction | | | | | |
|---|---|---|---|---|---|---|---|---|---|
| | Ref.39 | Ref.44 | Ref.45 | Landaas[38] | Deeth[39] | Deeth[44] | Deeth[45] | BLM[41] | BLM[45] |
| $C_2$ | 0 | | 0 | | 12 | | 12 | 15 | 16 |
| $C_3$ | 0 | | 1 | | 20 | | 19 | 22 | 24 |
| $C_4$ | 2 | | 2 | 24 | 34 | | 32 | 35 | 38 |
| $C_6$ | 4 | 0 | 5 | | 57 | 73 | 54 | 61 | 59 |
| $C_8$ | 14 | | 12 | | 73 | | 62 | 72 | 64 |
| $C_{10}$ | 42 | 46 | 27 | | 85 | 80 | 76 | 92 | 72 |
| $C_{12}$ | 71 | 65 | 50 | | 90 | 91 | 80 | 90 | 81 |
| $C_{14}$ | 78 | | 65 | | 90 | | 85 | 92 | 88 |
| $C_{16}$ | 80 | | 69 | 95 | 99 | | 95 | 95 | 91 |
| $C_{18}$ | 81 | 71 | 73 | 101 | 102 | 89 | 99 | 98 | 96 |
| $C_{18:1}$ | 77 | 80 | 68 | 96 | 97 | 88 | 97 | 96 | 96 |

Many of the extraction solutions prescribed are based on the heptane/isopropanol/sulphuric acid mixture originally proposed by Dole and Meinertz[43]. Because of the acidity of the solution the extraction of the lower fatty acids are much higher (though not complete) compared to the BDI-method. (See Table 1).
Citrate will not interfere but some lactate will be extracted if the milk samples are not fresh ($^{O}$SH>8). These methods are therefore not suited for sour or cultured milks (correction of the values might be possible). The great advantage of these methods is that the extraction is simple and rapid. The results obtained by the different methods    correlate well. One investigation[45] resulted in correlation coefficients of about 0.98 between BDI, Deeth and BLM method.

There have been devised extraction methods which give an almost complete recovery of lower fatty acids and no interference from lactate[40]. This is accomplished by using HCl/ether, centrifuging, transfer to another sample bottle, washing with HCl and treatment with silicic acid. Clearly this scheme is a little too involved for a routine method. The procedure is, however, very useful for analysis of sour and cultured products.

I find that a method based on the procedures of Deeth et al.[39], Landaas and Solberg[38] or the BLM-PEDIA[41,42] is useful for routine purposes. These procedures are rapid and give reproducible results. An outline of the methods is simple: Samples of milk are shaken with extraction liquid. An aliquot of the upper solvent layer containing the FFA is transfered to a titration-vessel and titrated with dilute ethanolic-KOH. Thymol blue or naphtholphthalein are used as indicator. Blank and preferably a standard (palmitic acid) must be included. Precise titration can be somewhat difficult and demands some experience as quite small amounts of FFA are to be determined.

An automated method. In Norway a large part of the FFA-analyses performed is centralized in our laboratory. The milk samples are collected in different parts of the country and sent to us. The samples originate from many different investigations: dairies check their farm milk, advisers monitor problem herds, new milking equipment is tested etc. This calls for an integrated scheme for sampling, transportation of samples and analysis: when samples are

taken they must immediately be stabilized against further lipolysis during storage and transportation; and at the laboratory
an automated and reliable analytical system capable of accepting
a large number of samples must be available.  In the last decade
such methods have been developed[46,47,48].  A system modified from
these methods has been successfully in operation in our laboratory
in the last few years[49].

We supply containers with sample tubes, sampling syringe and instructions.  The sample tubes contain 5 ml of an extraction liquid
composed of heptane/isopropanol/$H_2SO_4$[46].  When 1.5 ml milk are
added and the tube content vigorously shaken, the FFA and lipids
are extracted into an upper solvent phase.  These samples may be
stored at room temp. for up to four weeks without any significant
change in FFA content.  In the laboratory the tubes are shaken and
placed in the sampler of an Auto-Analyzer II.  The setup is similar
to that proposed by Suhren et al.[48].

Some points in the procedure are of importance to obtain reliable
operation and reproducible results:
1. Every tenth sample should be a standard (1 meq/l).  Some drift
   and carry-over can occur during long sample series and this
   can be corrected for by monitoring the standards (A computer
   system can do this automatically).  The method can be linearly
   standardized from 0 to 2.5 meq/l.
2. The system must be flushed with $N_2$ as $CO_2$ in air can cause
   serious interferences.
3. A reference channel is necessary to minimize drift.
4. The mixing coil for the sample channel must be made of Teflon
   (Kel-F).  Glass and PVC-tubing should be kept to a minimum.
5. Tubes and Kel-F coil in the sample channel must be changed
   frequently to minimize carry-over (Some components in the
   sample stream adsorb on the tube walls).
6. Long wash time between samples is advisable (We use 58 s
   sample, 32 s  wash).

As mentioned previously lactate can be extracted together with
FFA and interfere with the analyses.  We have investigated this
further and found that about 1.6 % of the lactic acid in milk is
extracted and that the acidity of samples should not exceed 8°SH.
(ml 0.25N/100 ml).  As samples are stabilized this seldom poses any

problems if fresh milks are sampled.

The extraction of FFA with this method is fast. A few seconds intensive shaking is enough to reach maximum extraction[47]. The extraction is, however, not complete for all FFA. The recovery is comparable with those reported for the method of Deeth et al.[39] (Table 1).

Typical values for reproducibility of the method are presented en Table 2[49]. These values are comparable with values obtained in Germany[48] (VC=4.35 % n=359) and Ireland[50] (VC=3.9 % n=137).

Table 2 Typical variation coefficients obtained (duplicates) by FFA-determination with the extraction/Autoanalyzer II method[49].

| FFA-level | n | $\bar{x}$ | S | VC (%) |
|---|---|---|---|---|
| < 0.5 | 14 | 0.443 | 0.0282 | 6.37 |
| 0.5-1.0 | 6 | 0.705 | 0.0341 | 4.84 |
| 1.0-1.5 | 15 | 1.235 | 0.0341 | 2.76 |
| 1.5-2.0 | 5 | 1.828 | 0.0798 | 4.37 |
| Total | 40 | 0.952 | 0.0386 | 4.05 |

The correlation with results from the BDI method is good for fresh farm milk[46,50] with r=0.94-0.98. A weaker correlation (r=0.74) was found when analyzing aged pasteurized bulk milk[50].

A reference/research method. A method to quantify the different FFA in milk must include two steps: quantitative isolation including removal from milk fat and determination of the concentrations of the different FFA. The first isolation step is almost impossible to perform by simple solvent partitioning: a liquid-solid partitioning is necessary for the removal of fat. Several methods have been proposed, the most common employing silicic acid/KOH[51,52,53,54] or anion exchange resins[55,56] as solid supports. The main drawback with these procedures is that fat hydrolysis may occur resulting in overestimation of FFA[57]. They also often require tedious pretreatment. Recently two attractive methods which over-

come these problems have been proposed[58,59]. They include the
use of inactivated alumina or the basic anion exhanger Amberlyst
26. Both methods start with an extraction of FFA/lipids with cold
HCl/diethyl ether (containing internal standards). This extraction
gives almost complete recovery of FFA[40]. To isolate FFA from
lipids the extract is either applied on a small alumina column
or shaken with Amberlyst 26. The dried alumina or Amberlyst 26
will have adsorbed the FFA nearly quantitatively. The FFA are
released from the support in two different ways. The important
advantage is that this does not include concentration or evaporation of extract, thereby minimizing possible loss. Gas chromatography is used to quantify the different FFA. They are released
from the Amberlyst 26 by methylation with HCl/methanol in a vial.
After addition of ether and saturated NaCl, samples are injected
directly into the gas cromatograph. As excellent GC-columns for
the separation of FFA now exist it is not really necessary to
methylate. In the method employing alumina the FFA are simply
released by addition of diisopropyl ether/formic acid. The supernatant is directly injected into the gas chromatograph equipped
with 10 % SP-216-PS columns. Concentration of FFA is calculated
with reference to the internal standards added during the extraction
step. The recoveries obtained by the two methods are good and have
been reported (Table 3).

Table 3. Reported recoveries of two adsorption/gas cromatographic
methods for the determination of individual FFA in milk.

| Fatty acid | % Recovery | |
|---|---|---|
| | Ref.58 | Ref.59 |
| $C_4$ | 92 | 98 |
| $C_6$ | 98 | 92 |
| $C_8$ | 95 | 94 |
| $C_{10}$ | 100 | 90 |
| $C_{12}$ | 100 | 98 |
| $C_{14}$ | 99 | 91 |
| $C_{16}$ | 107 | 91 |
| $C_{18}$ | 103 | 95 |
| $C_{18:1}$ | 104 | 91 |
| $C_{18:2}$ | | 99 |

The procedures outlined above seem to overcome the problems
of earlier methods, namely the loss of short-chain acids and
hydrolysis of lipids. They thereby offer a possibility for a
closer mapping of the FFA released during lipolysis.

## Conclusion, future

As I have tried to outline above, the methodology for the determination of FFA in milk has now come of age. It should now be
possible to select standard procedures with defined fields of
application. The FFA methods I have mentioned give results in
meq/l which I find convenient. The correlation between the methods
has been (at last partly) established.

Future research in analytical methodology for the study of lipolysis should maybe put more emphasis on the determination of other
factors involved. Some work has been done towards this. Several
methods have been devised for the determination of "free fat"
[60,61,62] as this fat was thought more susceptible to lipase attack.
Some workers[63] have found a correlation between free fat and
lipolysis. The usual heating/centrifuging and solvent extraction
methods for free fat are, however, usually not useful in this
context[64]. It has been stated[64] -"that no single index can be
used adequately to describe the effects of agitation on the milk
fat globule as it relates to the extent of lipolysis by milk
lipase". Methods employing Candida lipase do, however, show some
promise.

Use of enzyme immunoassay can be of value in the study of lipase
action. It has been used with good results to measure the relative
amount of lipoprotein lipase activator in milk[11].

It is to be hoped that these and other novel analytical methods
will contribute more information about cause and subsequent risk
for lipolysis than a measurement of FFA level.

## References

1. International Dairy Federation, 1975, Bulletin
   (Document no. 86).
2. H.C.Deeth and C.H.Fitz-Gerald, Austr.J.Dairy Techn.,1976,**31**,53.
3. M.G.Fleming, Ir.J.Food Sci.Technol., 1979,**3**,111.

4. International Dairy Federation, 1980, Bulletin (Document no.118).

5. H.B.Castberg, T.Egelrud, P.Solberg and T.Olivecrona,
   J.Dairy Research, 1975,42,255.

6. H.B.Castberg and P.Solberg, Meieriposten, 1974,63,961.

7. M.Anderson and E.C.Needs, J.Dairy Research, 1983,50,309.

8. H.C.Deeth and C.H.Fitz-Gerald, J.Dairy Research, 1977,44,569.

9. F.M.Driessen and J.Stadhouders, Neth. Milk Dairy J. 1974,28,130.

10. A.Jellema, Neth.Milk Dairy J., 1975,29,145.

11. M.Anderson, J.Dairy Science, 1979,62,1380.

12. R.A.Clegg, J.Dairy Research, 1980,47,61.

13. G.Sundheim, T.-L.Zimmer and H.N.Astrup, J.Dairy Science, 1983,
    66,400.

14. G.Sundheim, T.-L.Zimmer and H.N.Astrup, J.Dairy Science, 1983,
    66,407.

15. A.M.A. Salih and M.Anderson, J.Dairy Research, 1979,46,623.

16. R.Gudding, J.Food Protection, 1982,45,1143.

17. M.E.Jurczak and A.Sciubisz, Milchwissenschaft, 1981,36,217.

18. K.C.Bachman, J.Dairy Science, 1982,65,907.

19. H.N.Astrup, P.Skrøvseth, L.Vik-Mo, A.Ekern and E.Sola,
    Meieriposten, 1977,66,681.

20. H.N.Astrup, L.Vik-Mo, O.Skrøvseth and A.Ekern, Milchwissen-
    schaft, 1980,35,1.

21. H.N.Astrup, L.Bævre, L.Vik-Mo and A.Ekern, J.Dairy Research,
    1980,47,287.

22. M.Anderson, J.Dairy Science, 1974,57,399.

23. P.Nilsson and S.Willart, Milk and Dairy Research, Alnarp,
    Sweden, 1961, Report No.60.

24. W.K.Downey, International Dairy Federation, Bulletin, 1980,
    Doc. 118,4.

25. N.P.Tarassuk and J.L.Henderson, J.Dairy Science, 1942,25,801.

26. M.Anderson, J.Dairy Research, 1981,48,147.

27. P.Nilson-Ehle, T.Egelrud, P.Belfrage, T.Olivecrona and
    B.Borgstrøm, J.Biol.Chemistry, 1973,148,6734.

28. T.Olivecrona, International Dairy Federation, Bulletin,1980,
    Doc.118,19.

29. B.H.Webb, A.H.Johnson and J.A.Alford, ed. "Fundamentals of
    Dairy Chemistry", The Avi Publishing Company, Inc.,Westport,
    Conn., (Second Ed.1974), Chapter 4, pp. 153-164.

30. W.F.Shipe, G.F.Senyk and K.B.Fountain, J.Dairy Science,1980,
    63,193.

31. V.T.Pillay, A.N.Myhr and J.I.Gray, J.Dairy Science, 1980, 63,1213.

32. S.Kuzdzal-Savoie, International Dairy Federation, Bulletin, 1980, Doc.118,53.

33. E.L.Thomas, A.J.Nielson and J.C.Olson jr.,Amer.Milk.Rev.1955, 77,50.

34. Statens Forsøgsmejeri, Hillerød, Denmark,1962,136.Beretning.

35. K.Singsaas and G.Hadland, Meieriposten, 1972,61,153.

36. F.M.Driessen, A.Jellema, F.J.P.van Luin, J.Stadhouders and G.J.M. Wolbers, Neth.Milk Dairy, 1977,31,40.

37. E.N.Frankel and N.P.Tarassuk, J.Dairy Science, 1955,38,751.

38. A.Landaas and P.Solberg, Meieriposten, 1974,63,497.

39. H.C.Deeth, C.A.Fitz-Gerald and A.F.Wood, Aust.J.Dairy Techn., 1975,30,109.

40. A.M.A. Salih, M.Anderson and B.Tuckley, J.Dairy Research,1977, 44,601.

41. L.Mouillet, F.M.Luquet, H.Nicod, J.F.Boudier and H.Mahieu, Lait, 1981,61,171.

42. J.Thomasow, Deutsche Molkerei-Zeitung, 1981,28,906.

43. V.P.Dole and H.Meinertz, J.Biol.Chemistry, 1960,125,1595.

44. H.Hänni and M.Rychener, Mitt.Gebiete Lebensm.Hyg., 1980,71,509.

45. J.Van Crombrugge, R.Bossuyt and R.Van Renterghem, Revue de l'Agriculture, 1982,35,3313.

46. B.Lindqvist, R.Roos and H.Fujita, Milchwissenschaft,1975,30,12.

47. Statens Forsøgsmejeri,Hillerød, Denmark,1975,210.Beretning.

48. G.Suhren, W.Heeschen and A.Tolle, Milchwissenschaft,1977,32,641.

49. G.Bråthen, Meieriposten, 1980,70,345.

50. T.C.A. Mc Hann, D.C.Eason and W.K.Downey, International Dairy Federation, Bulletin, 1975, Doc.No 86,184.

51. W.J.Harper, D.P.Schwartz and I.S.El-Hagaraway, J.Dairy Science, 1956,39,46.

52. M.Iyer, T.Richardson, C.H.Amundsen and A.Boudreau, J.Dairy Science, 1967,50,285.

53. A.H.Woo and R.C.Lindsay, J.Dairy Science, 1980,63,1058.

54. A.H.Woo and R.C.Lindsay,J.Dairy Science, 1982,65,1102.

55. J.A.Kintner and E.A.Day, J.Dairy Science, 1965,48,1575.

56. D.D.Bills, I.I.Khatri and E.A.Day,J.Dairy Science,1963,46,1342.

57. W.Stark, G.Urbach and J.S.Hamilton, J.Dairy Research,1976,43, 469.

58. H.C.Deeth, C.H.Fitz-Gerald and A.J.Snow, New Zealand J.Dairy Science and Techn.,1983,18,13.

59. E.C.Needs, G.D.Ford, A.J.Owen, B.Tuckley and M.Anderson, J.Dairy Research, 1982,50,321.

60. A.Dillier-Zulauf, A. and P.R.Wirasekara, Deutsche Molkerei-zeitung, 1971,92,1943.

61. I.E.Te Whaiti and T.F.Fryer, New Zealand J.Dairy Science and Techn., 1975,10,2.

62. O.Aule and H.Worstorff,International Dairy Federation,Bulletin, 1975, Doc.no.86,116.

63. I.E. Te Whaiti and T.F.Fryer, New Zealand J.Dairy Science and and Techn. 1976,11,273.

64. H.C.Deeth and C.H.Fitz-Gerald, J.Dairy Research,1978,45,373.

# Determination and Occurrence of Organophosphorus Pesticide Residues in Milk

CENTRAL LABORATORY FOR QUALITY ASSURANCE OF NESTLÉ PRODUCTS TECHNICAL
ASSISTANCE CO. LTD., CH-1814 LA TOUR DE PELIZ, SWITZERLAND

## Introduction

The use of pesticides directly on livestock, especially on
dairy cows, for protection against disease vectors such as
mites, ticks and insects is widespread. Formerly, organochlorine pesticides were the most commonly used, but the high residues in milk and meat associated with their use lead to their
gradual replacement by specific organophosphorus compounds
which are more rapidly metabolised.

It is usually assumed that these compounds occur only in minute
quantities in milk, but little is known about the levels
actually found in practice.

Analytical methods available for the analysis of organophosphorus pesticide residues deal mostly with the estimation of a
single compound and cannot necessarily be applied to regulatory
work. The analyst who wants to check a sample of milk for
conformity to the Codex maximum residue limits needs a multi-
residue procedure that allows the qualitative and quantitative
determination of low levels of a whole series of compounds and
this, preferably with a minimum of extraction and clean-up steps.

There are some published multi-residue methods that describe
the simultaneous determination of organochlorine and organophosphorus pesticides (OPs) in fatty substrates (1,2,3), but
these procedures have several drawbacks.

The well-known AOAC procedure (1) is rather laborious and its
suitability for determining OPs in milk and dairy products
has been tested for very few compounds. Because of its rapidity the method of Stijve and Cardinale (2) is frequently used
for monitoring levels of organochlorine pesticides in fresh
milk. It allows determination of some non-polar OPs, such as
bromophos and chlorpyrifos, but is not suitable for determining
medium polar or polar compounds which co-elute from the Florisil
column with fats. The multi-residue method of Specht (3) allows
the simultaneous determination of about 85 pesticide compounds
including 30 OPs in non-fatty and fatty substrates, but few
recovery experiments have actually been performed on milk and
milk products. Furthermore, the procedure uses automated
gel permeation clean-up which is not readily accessible to
smaller residue laboratories.

In this laboratory we were confronted with the need to extend
our monitoring programme, which was limited to organochlorine
pesticides, to 20 OPs ranging from the water-soluble dicrotophos to such non-polar compounds as fenchlorphos and fenitro-
thion. None of the above-mentioned methods alone was entirely
suitable for that purpose, but the problem could be tackled by
using a combination of existing techniques.

A detailed description of the method thus obtained, and the
experience gathered with OP residues in milk over a period of
10 years are communicated in this paper.

## Experimental

1.          Principle of the method

            Extraction with chloroform. Evaporation of solvent.
            Isolation of very polar OPs (dimethoate and dicroto-
            phos) by elution from a Celite column using an
            aliquot of the fat. Clean-up of medium polar and
            non-polar pesticides by acetonitrile partitioning
            combined with a "back wash" operation of a second
            aliquot of fat dissolved in light petroleum. Dilution
            of combined acetonitrile fractions with water,
            followed by partitioning of the pesticides into light
            petroleum-dichloromethane 4:1 v/v. Gas chromatographic
            (GLC) determination of the OP compounds in the crude
            concentrated extract. Clean-up on a small Florisil
            column and estimation of organochlorine pesticides
            and PCBs by electron capture GLC. Confirmation of
            identity by thin-layer chromatography.
            The method can be applied to fresh and condensed milk,
            milk powder, infant foods on milk basis, butter and
            cheese.

2.          Chemicals and apparatus

            - Light petroleum, acetonitrile, methylene chloride
              and ethyl acetate, quality for residue analysis

            - Methanol, ethanol, acetone, chloroform, iso-octane,
              the GR quality of Merck is adequate

            - Anhydrous sodium sulfate GR and Celite 545, refined
              diatomaceous earth. Remove interfering impurities
              by heating overnight at 550 $^\circ$C

            - Florisil, synthetic magnesium silicate, 60-100 mesh.
              Remove impurities and standardize as described by
              Stijve and Cardinale (ref. 2).

            - Gas chromatographs, simple one column- one detector-
              instruments are preferable to those having multi-
              detector and dual column arrangements.
              The electron capture detector should have picogram
              sensitivity for organochlorine compounds such as

lindane and extended linear response.
The phosphorus detector can either be of the flame
photometric or thermionic type.

- Pre-coated plates for TLC analysis as described in
ref. 2.

- Rotavapor apparatus - rotative evaporator connected
to a waterjet pump.

- Various laboratory glassware including chromatography tubes fitted with stopcocks 8 x 200 mm,
having a 100 ml reservoir at the upper end.

3.      Procedure

3.1     Preparation of the test portion

Weigh 50 g homogenized milk in a 1 lt conical flask.
When analysing evaporated and sweetened condensed
milk take 25 g and add 25 ml water.
Dissolve powdered milk products by vigorously stirring
10 g with 30 ml water. Transfer lump-free solution
to a 1 lt flask using 2 portions of 10 ml water as
rinses.
Heat butter to about 60 °C until the fat separates
and decant through a plug of glass wool.
Extract fat from cheese with methylene chloride-
methanol as outlined in ref. 2.
Subject fat from butter and cheese to solvent partition
(3.4.1)

3.2     Extraction

Add 150 ml of chloroform to the test portions of milk
and reconstituted milk as obtained under 3.1.
Stopper the flask and shake vigorously during two
minutes. Subsequently, add slowly 150 g of anhydrous
sodium sulfate, swirling the flask after each addition.
Allow the solids to settle during 10 minutes. If
the supernatant liquid is not clear, add an additional
portion of 25 g of sodium sulfate. Decant the clear
extract through a plug of glass wool into a 1 lt
round bottomed flask.
Extract the milk-sodium sulfate mixture twice more with
100 ml of chloroform, shaking vigorously and allowing
solids to settle each time, before pouring the extracts
into the round bottomed flask. Evaporate in a rotavapor
apparatus under reduced pressure at a temperature of
about 40 °C until an oily residue remains. Remove the
last traces of chloroform by adding 50 ml of light
petroleum and evaporating again.

Take up oily residue in a few ml of light petroleum
and transfer quantitatively to a 10 ml volumetric
flask. Make up to volume with the same solvent and

shake well to mix. If residues of all possible present
compounds, including dicrotophos and dimethoate, are
to be determined, take 5 ml for the multiresidue determination and start with the solvent partition as described under 3.4.2.
Use remaining 5 ml for clean-up outlined for polar
OP residues (3.3). If need be, the whole extract may
be taken for either kind of analysis.

3.3.        Clean-up for dicrotophos and dimethoate residues

3.3.1       Preparation of the chromatography column

Weigh 10 g Celite 545 into a 100 ml glass beaker, add
exactly 3 ml water and mix rapidly with a glass rod
until a homogeneous free flowing powder is obtained.
Pack into a chromatography tube dry, from bottom to top
with tapping, a plug of glass wool, 0,5 g of dry Celite
545, the 13 g Celite-water mixture (added in about
5 portions with each packed uniformly and tightly),
again 0,5 g of dry Celite and, finally, 10 g of anhydrous sodium sulfate. Prewash the column with 50 ml
of light petroleum and discard the washing.

3.3.2       Selective elution of dicrotophos and dimethoate

Bring the 5 or 10 ml extract, obtained under 3.2
quantitatively on the top of the column using a few ml
of light petroleum as a rinse. After the extract has
entered the column, elute with the following mixtures :

1.   50 ml of light petroleum
2.   50 ml of light petroleum-methylene chloride 4:1 v/v
3.   70 ml of light petroleum-methylene chloride 2:1 v/v

Discard the first fraction which contains fat and other
interferences. Gather the second and third eluates
which contain respectively dimethoate and dicrotophos
in 250 ml round bottomed flasks. Evaporate in a rotavapor apparatus until about 5 ml are left and remove the
rest by blowing with a stream of clean air. Rinse the
flasks repeatedly with small portions of ethyl acetate.
Transfer washings to 5 ml volumetric flasks, make up to
volume and shake well to mix.

3.4         Solvent partition

3.4.1       Butter and cheese fat

Weigh 3 g in a 125 ml separating funnel.
Add light petroleum so that total volume of fat and
solvent is 15 ml. Add 30 ml acetonitrile saturated
with light petroleum and shake vigorously during 1 min.
Let layers separate and drain acetonitrile into a second
separating funnel containing 15 ml of light petroleum.
Shake vigorously during 1 minute, let layers separate
and transfer acetonitrile phase to a 1 lt separating

funnel containing 600 ml of distilled water.
Pass the acetonitrile phase from each of three additional partitionings through the same 15 ml of light
petroleum in the second separating funnel. Shake vigorously each time and combine acetonitrile extracts in
the 1 lt separating funnel.
Add 100 ml of light petroleum-methylene chloride mixture 4 : 1 v/v to the cloudy acetonitrile-water
dilution. Shake vigorously during 1 minute and allow
the layers to separate. Discard aqueous layer and wash
extract twice with 100 ml portions of water. Discard
washings. Dry extract with anhydrous sodium sulphate
and transfer it to a 500 ml round bottomed flask. Wash
sodium sulphate with small portions of light petroleum
and add washings to the extract.
Evaporate the extract in the rotavapor apparatus until
a few ml are left. Remove the remaining solvent by
gently blowing with an air current. Take up residue in
about 2 ml of light petroleum and transfer quantitatively with small portions of the same solvent to a 10 ml
volumetric flask. Make up to volume and shake well to
mix. This final extract should only contain traces of
fat.

3.4.2     Other milk products

Transfer appropriate volume of test portion extracts as
obtained under 3.2 to a 125 ml separating funnel.
Adjust volume to 15 ml and perform acetonitrile
partition as described above.

3.5       Clean-up for organochlorine pesticides

Subject an aliquot of extract to Florisil clean-up as
outlined in ref. 2 or 4.

3.6       Gas chromatography (GLC)

3.6.1     Organophosphorus pesticides

Multiresidue analysis of OPs is generally performed at
10 percent silicone DC-200 as specified in the AOAC
method ( 1 ). However, this column is not able to
separate critical pairs such as methylparathion and
chlorpyrifos methyl, and malathion and pyrimiphos
methyl.
If these pesticides are to be determined, it is recommended to take a 180 x 3 mm column filled with 5 % Dexsil
300.
Volatile pesticides such as dicrotophos, dichlorvos
and dimethoate should be chromatographed at a column
temperature of about 170 $^\circ$C, higher boiling compounds
e.g. coumaphos at 240 $^\circ$C. Detector response for some
labile OPs can be much improved by priming the column
with mcg amounts of the compound.

For accurate quantitation it is necessary to have both
sample extract and reference compounds in the same solvent, preferably ethyl acetate.

3.6.2        Organochlorine pesticides and PCBs

Perform GLC analysis as outlined in ref. 2.

3.7.         Thin-layer chromatography (TLC)

Confirm GLC results for OPs by TLC of extracts obtained
under 3.3.2, 3.4.1 and 3.4.2, using mobile phases and
enzymic detection procedure as given in ref. 5.
For organochlorine pesticides use TLC on alumina layers
and photochemical detection with silver nitrate (ref. 2).

4.           Recoveries

In a series of recovery experiments at the 5 and
50 mcg/kg level the following results were obtained.

| | | | |
|---|---|---|---|
| Dichlorvos | 64-72 | Coumaphos | 87-92 |
| Dicrotophos | 72-96 | Dioxathion | 80-87 |
| Dimethoate | 71-90 | Diazinon | 91-95 |
| Bromophos | 83-93 | Chlorfenvinphos | 81-96 |
| Iodofenphos | 90-98 | Crufomate | 92-105 |
| Chlorpyrifos | 92-97 | Malathion | 88-102 |
| Chlorpyrifos methyl | 84-89 | | |
| Fenitrothion | 81-83 | Ethion | 80-83 |
| | | Pirimiphos methyl | 87-95 |
| Phosalone | 92-97 | | |
| Quintiophos | 87-89 | | |

Recoveries for organochlorine pesticides tend to be
equal to those obtained with the direct elution methods
( 2,4 ). For essential non-polar compounds such as HCB
and PCB slight losses occur during the "back wash" step,
but recoveries are still at or above 80 percent.


Occurrence of organophosphorus pesticides in milk

During the last ten years we analysed 1780 samples of milk for
both organochlorine and organophosphorus pesticides residues.
The material consisted of milk samples from individual cows,
separate farms, bulk milk from various milk routes, milk processing plants and the bottled and standardised product available
on markets in Europe and two African countries.
All samples contained organochlorine pesticides in concentrations
ranging from a few mcg/kg to 10 mg/kg on fat basis. HCH isomers
and DDT plus its breakdown products were most frequently encountered. HCB and heptachlor epoxide were absent in milk from
Africa and from Brazil. Milk from Africa frequently contained
camphechlor residues.

In Europe, concentrations exceeding the Codex maximum residue
limits were rather exceptional, but milk in Africa was often
grossly polluted with camphechlor, introduced by the cattle
dipping practice.

Table I lists the OPs found and their respective concentrations.
In all, only 86 samples (4,8 %) contained detectable amounts of
these insecticides and only 9 compounds were involved. The number
of positives should be interpreted with caution, because in a few
instances samples were taken because it was known that cattle
dipping with a certain compound was going on in a specific area.
For example, the data for dioxathion and coumaphos belong in this
category. In each case a dozen samples were taken which were all
positive. On the other hand, the other OPs have been found as
occasional positives in large series of samples.
Bromophos, chlorpyrifos, diazinon, iodofenphos and fenchlorphos
were only found in European milk. Contamination with chlorpyrifos
seems to be on the rise, not only of milk but of various other
produce.
This can be explained by a certain over-enthusiastic use of this
insecticide in factories, where it is frequently used against
cockroaches. We first observed this compound repeatedly when
analysing milk powders from a certain factory. Upon investigating
the source of the contamination, it was found that the factory
premises were treated regularly with a chlorpyrifos formula.
This was already strongly suggested by the fact that the insecticide could be extracted by simply shaking the milk powder with
non-polar solvents such as hexane or toluene. Apparently, chlorpyrifos adhered loosely to the powder particles, and was not
incorporated in the fat phase as would have been the case with
an incurred and metabolised residue. Sometimes, chlorpyrifos was
accompanied by diazinon, probably because some commercial formulas contain both insecticides.

Coumaphos, an almost obsolete acaricide, was exclusively found
in milk from a certain region in Brazil. Chlorfenvinphos
occurred in milk from Africa and was not seldom accompanied by
considerable quantities of camphechlor, suggesting that the two
compounds are formulated together.
Quintiophos or Bacdip is also used for cattle dipping in Africa.
Although we repeatedly analysed milk of cows within a few days
after treatment, we were never able to detect any residues. The
limit of detection of the method was as low as 2 ppb on milk.

Stability of two organophosphorus pesticides
during milk processing.
---

Some experiments were carried out to measure the stability of
two recommended acaricides, one fat-soluble and one water-soluble
compound, delnav (dioxathion) and dicrotophos (bidrin), under
conditions of milk processing. For this purpose, known amounts
of both compounds were dispersed in fresh milk which was subsequently used for the manufacture of experimental batches of
sweetened condensed milk and spray dried milk powder. Determinations were carried out at each step of the manufacturing process. The data reported in table II indicate that both delnav

TABLE I          Occurrence of organophosphorus
                 pesticide residues in milk samples (N=1780)

| Compound | Number of positive samples | Range in mg/kg on fat basis |
|----------|----------------------------|-----------------------------|
| Dioxathion | 12 | 0,2   -  0,7 |
| Bromophos | 15 | 0,08  -  1,25 |
| Chlorfenvinphos | 9 | 0,01  -  0,15 |
| Dicrotophos | 7 | 0,005 -  0,02  * |
| Coumaphos | 12 | 0,1   -  0,48 |
| Chlorpyrifos | 21 | 0,005 -  0,11 |
| Diazinon | 5 | 0,01  -  0,15 |
| Iodofenphos | 2 | 0,01  -  0,03 |
| Fenchlorphos | 3 | 0,03  -  0,1 |
| TOTAL | 86 = 4,8 % | * expressed on whole milk |

TABLE II     Fate of dicrotophos (bidrin) and dioxathion
             during manufacture of experimental batches
             sweetened condensed milk and milk powder

Sweetened condensed milk

|  | Quantity in mg/kg of | |
|--|------------|-----------|
|  | dicrotophos | dioxathion |
| Fresh milk | not detectable | not detectable |
| Fresh milk after addition of 0,2 mg/kg | 0,20 | 0,20 |
| After pasteurisation | 0,20 | 0,18 |
| After condensation with sugar (final product) | 0,22 | 0,40 (0,50) |

Spray dried milk powder

|  | dicrotophos | dioxathion |
|--|------------|-----------|
| Fresh milk | not detectable | not detectable |
| Fresh milk after addition of 0,2 mg/kg | 0,22 | 0,17 |
| After pasteurisation | 0,21 | 0,17 |
| Precondensate | 0,85 | 0,50 |
| After spray-drying (final product) | 1,2 | 0,95 (1,60) |

Figures in brackets indicate theoretical values of pesticide
compounds in final product, making allowance for the concentration
factor.

and dicrotophos are stable, although about 50 percent of the
latter was lost during the manufacture of sweetened condensed
milk. Spray drying had little effect on both compounds.

## Discussion

The present analytical method for the multiresidue determination
in milk and milk products is a combination of existing techniques:
chloroform extracts all compounds in good yield provided that
the operation is performed in presence of water. The fatty
residue obtained after evaporation of the solvent is further
cleaned up in two ways. Polar pesticides are recovered from a
Celite column ( 6 ), whereas medium polar and non-polar compounds are isolated by the classic acetonitrile partition
procedure involving a "backwash" step with light petroleum to
minimize presence of lipids in final extracts ( 7 ).

Without further clean-up OPs are determined by GLC with a
thermionic or flame photometric detector, whereas organochlorine
pesticides are estimated by electron capture GLC after clean-up
on a Florisil column. For the sake of rapidity and economy this
last step can be miniaturized ( 4 ).

The method is adequate for all organochlorine pesticides normally
encountered in milk, and yields good recoveries for 20 OPs of
various polarity, which suggests that it can probably be used for
the determination of residues of many other compounds, including
a.o. organonitrogen pesticides. Preliminary results obtained in
this laboratory suggest that the procedure also recovers synthetic pyrethroids and such simple carbamates as carbaryl and
propoxur.

Attention should be drawn to the fact that several non-polar OPs
appear in the first eluate of the Florisil column; analysts not
familiar with this phenomenon have mistaken such OPs for
organochlorine compounds.
On several packed stationary phases used for monitoring OC
residues by GLC bromophos has the same retention time as heptachlor epoxide. This has caused embarrassment to a few official
food inspection laboratories who objected to the latter compound
in milk products because its concentration was said to exceed the
Codex MRL of 0,15 mg/kg on fat basis.
Investigation proved that the products contained virtually no
heptachlor epoxide. The peak interpreted as such was in reality
bromophos of which the electron capture response is approximately
5 times lower.

Thus, the bromophos concentration in the products was 5 times the
measured quantity of heptachlor epoxide. The most contaminated
sample contained 1 mg/kg on fat basis, which is still below the
limit for bromophos which is 0,05 mg on whole milk or 1,25 mg/kg
on fat basis for milk products.
Similar errors have occurred with iodofenphos and endrin, and,
more recently, with chlorpyrifos methyl residues in cereals,
which were mistaken for aldrin.

The experience gathered with analysing milk and milk products for
OPs confirms earlier observations ( 8 ) that these compounds do
not pose a serious residue problem. Indeed, few residues are
found and the concentrations are generally below the maximum
allowable limits. A notable exception, however, is dioxathion.
Residues associated with the cattle dipping practice almost inva-
riably exceed the Codex limit of 0,2 mg/kg on fat basis.
The limit was established without taking the said treatment into
account. Since there were no data available for dioxathion residues in milk, the Codex specialists decided that this foodstuff
was not supposed to contain any, and recommended a limit which
was at or about the limit of detection of the analytical method.
It would seem that a case exists for reconsidering the limit,
because dioxathion is a rather cheap and effective chemical for
the control of ticks which has advantageously replaced the use
of the far more objectionable camphechlor in a number of
countries.
Although almost all OPs are rapidly broken down during their
passage through the cow's metabolic system, the results of the
stability experiments reported in this paper underline once
more the need to keep pesticide formulas and sprayers away from
areas where milking equipment is located, since contamination
from utensils is unlikely to disappear during processing.

References

1. W. Horwitz (edit.) Official Methods of the Association
   of Official Analytical Chemists, Washington DC.
   13th Edition (1980). 29.001 - 29.018 - Multiresidue
   methods for chlorinated and certain organophosphorus
   pesticides.

2. T. Stijve and E. Cardinale, Mitt. Gebiete Lebensm. Hyg.,
   1974, 65, 131.

3. W. Specht and M. Tillkes, Z. anal. Chem., 1980, 301, 300.

4. T. Stijve and E. Brand, Deutsche Lebensm.-Rundschau, 1977,
   73, 41.

5. T. Stijve and E. Cardinale, Trav. chim. aliment., 1971, 62,
   24.

6. G. Zweig (edit.) Analytical Methods for Pesticides,
   Plant Growth Regulators, and Food Additives Academic Press,
   New York - San Francisco - London
   Volume V (1967), Chapter 9 : Bidrin insecticide, 213.

7. Pesticide Analytical Manual, as revised June 1979,
   Food and Drug Administration, Washington, DC, USA.
   Vol. I, Section 211.14b.

8. B. Marek in International Dairy Federation (IDF) Bulletin
   1979, 113, Chapter 2 : Organophosphates & Carbamates.

# Determination of Intensity of Maillard Reactions in Dairy Products

By M. Carić*, D. Gavarić, and S. Milanović
FACULTY OF TECHNOLOGY, DAIRY DEPARTMENT, UNIVERSITY OF NOVI SAD, 21000
NOVI SAD, VELJKA VLAHOVICA 2, YUGOSLAVIA

INTRODUCTION

Possible nonenzymatic browning reactions in dairy products. Nonenzymatic browning reactions in dairy products following high temperature treatment could be classified as:
a) Caramelazation, i.e. brown discoloration of sugar without
the presence of amino component;
b) Browning caused by Maillard reactions between lactose and amino
component;
c) Oxidative browning[1].

Caramelization is thermal degradation of lactose; it takes
place in the absence of amino component and needs high energy of
activation. As far as proteins are always present beside lactose in dairy products, and the process temperatures are relatively low for caramelization, the essential cause of browning in
heated dairy products are melanoidines, present as a result of
Maillard condensation between lactose and milk proteins. Oxidative browning develops rarely in dairy products.

On the basis of different authors' investigations, the
reactive way of Maillard browning is nowadays mainly known: from
the principal reactants - lactose and casein, to the final product - melanoidines. Amino acids in casein which react mostly are
the N - terminal residue of the polypeptide chain and basic amino acids, especially lysine[2]. Enol form and Schiff's base were
first identified in condensed unsweetened milk and in skim milk
powder, while, hydroxymetylfurfural was identified in condensed
unsweetened milk, and lactosin and fructosin in sterilized (UHT)
milk[3].

The organoleptic changes caused by Maillard reactions in
dairy products are considered as undesirable, and it is supposed
that they should be, by regular process control.

minimized. The exceptions are dairy products assigned for some
confectionary products. The intensity of changes caused by
Maillard reactions is, above all, defined by : severity of heat
treatment (temperature and time), and concentration of casein
and lactose during processing; and water content in product and
temperature and time during storage. Besides, the degree of heat
induced changes by Maillard reaction in dairy products is also
affected by redox potential, buffer capacity and pH-value of milk,
which are deffined by ions, gases and casein concentrations.
Analogous to the mentioned changes caused by Maillard reactions,
the nutritive and biological value of the product, drops, because
of the decreased availability of lysine, bond to lactose ( 4 - 75
%, depends on processing ) in human metabolism.

Processing and storage parameters cause arising of Maillard type reaction products to a different degree in particular
dairy product. Those primary are : concentrated and dried dairy
products, followed by sterilized milks, processed cheese, and less
expressed by some other dairy products. Milk powders and
sterilized milks were chosen for these investigations as products
with mostly expressed heat induced changes of Maillard type.

Maillard reactions by dried milks. As far as dry dairy products
are industrially produced exclusively by usage of heat, these
products are more or less followed by Maillard's reactions. The
degree of heat induced changes Maillard type is determined by
the way of drying and parameters of drying and storage. It is
found that the forewarming temperature affects significantly the
5-HMF content in the final product too[5]. Two stage drying, using
fluid-bed dryer after spray drying, leads to the increase of the
content of Maillard reactions products 0 - 10%, depending on the
time and temperature in fluid-bed dryer[6]. Intensification of Maillard's reactions by instantizing is understandable, taking in
account the usual way of instantizing process and the reactants
concentration influence on the rate of reaction. Both ways of instantizing ( straight through, and two stage ) means prolonged
drying in the moment of arised reactants concentrations ( 10-15%
$H_2O$ )[7].

Browning reactions of Maillard type with all associated
changes are intensified during storage of dry dairy products if
the moisture content is 5% or higher[8]. Products with lower moisture content do not undergo these changes, only if they are pro-

perly packaged.

Maillard reactions by UHT milks. Detailed study about lactose interaction with milk proteins as the result of UHT treatment with identification of proteins, involved, was carried out by Turner et al[2]. Results showed that it is essentially casein which reacts, or more exactly ᴋ-casein. The reason is not definitely clear, but the matter of fact is that protein – – protein interactions of heat denaturated whey proteins could possibly make the lysyl residues less available for the reaction with lactose[2]. The data about the interaction between β-lactoglobuline and lactose, in the absence of casein given by Ludvig [9,10] showed that the additional disulfide bridges are formed in macromolecular products of Maillard reaction which protect lysine, so that about 40% lysine is left intact. If the high temperature treatment is carried out after the reduction of intramolecular disulfide groups and blocking thiol groups in β-lactoglobuline, the protective effect is missing and whole lysine reacts in Maillard reaction.

Investigations carried out by Samuelsson[11] showed that the intensity of Maillard reactions is much stronger in milk sterilized by autoclaving than in UHT sterilized milk. Möller et al.[12] established that about 10-30% lysyl residues from casein in UHT milk stored 6 months to 3 years was involved for the lactulose-lysine formation, which afterwards turns to fructose-lysine. The later change takes place more quickly at higher storage temperatures. The interaction of ε-amino group of lysine with lactose changes the electrophoretic charac-teristics of casein[13] and its sensibility to chymosin[14].

Once started at high temperatures, Maillard's reactions exhibit autocatalytic qualities further on, even in refrigerated products, so they could cause the destabilization and gelation of UHT and concentrated milks during prolonged storage[13].

MATERIAL AND METHODS

Enriched milk powders were obtained by drying evaporated skim milk on laboratory "Anhydro" spray dryer with inlet air temperature $t_i$=220°C, and outlet air temperature $t_o$=90°C. Besi-des control, several different enriched samples were made:
1. Milk powder with $1 \times 10^{-2}$kg added iron / $m^3$ reconstituted milk, in the form of "Fortepan F", producer "Pliva", Zagreb,

Yugoslavia;

2. Milk powder with $5 \times 10^6$ IU added vitamin A/ $m^3$ reconstituted
   milk, in the form of $AD_2$ vitamin preparation, producer
   "Hoffman La Roche", Basle, Switzerland.
   Control and samples were stored at room temperature during 8 months and analyzed every 30 days.
   All sterilized enriched milks were produced on industrial
   scale in dairy "Standard" - PKB, Belgrade, on line for
   aseptic sterilization VTIS Alfa-Laval. The processing
   parameters were:
   milk preheating at $80^\circ C$, homogenization at $1.47 \times 10^7$ Pa,
   sterilization at $140^\circ C$ during 3-4 s. Besides control,several
   different enriched samples were made:

1. Milk with $1 \times 10^{-2}$ kg / $m^3$ added iron in the form of $FeSO_4 \times 7H_2O$;
2. Milk with 1% added sodium-caseinate, producer "Gervais Dano-
   ne", Rosenheim, W.Germany;
3. Milk with $5 \times 10^6$ IU / $m^3$ added vitamin A, producer "Hoffman
   La Roche", Basle, Switzerland;
4. Sterilized milk enriched simultaneously with above components in mentioned levels.

Samples were at room temperature during 3 months and
were analyzed every 15 days.

Content of 5-hydroxymethylfurfural (5-HMF) in all samples
was established by Keeney and Bassette[7] method.

RESULTS

Milk powders. The content of 5-HMF in enriched milk powders and
control is shown in Table 1. As it was expected 5-HMF level in
milk powder was much higher compared with sterilized milk. In
control milk powder 5-HMF content was lower again, than in enriched powders. Four days after production the content in control
was 5.68 mmol/$m^3$, what is in good agreement with literature
data[15]. The content was higher in milk powder with added iron,
but the highest in the powder with added vitamin A. Thirty days
later this content is more than doubled in control sample,with
steady increase during storage. Two month after production in
milk powder with added iron 5-HMF content almost tripled showing
also steady increase during storage. With powder enriched with
vitamin A the same happened at the end of storage period. But,
generaly, there was tendency in 5-HMF to equalize in all samples,
what was observed 5 months and later after production.

Table 1.: Changes in 5-HMF content in enriched milk powders during storage. Samples: 1-Milk powder with added iron, 2-Milk powder with added vitamin A, 3-Control.

Content of 5-HMF ($mmol/m^3$)

| Sample No | Storage period (days) | | | | | | | | |
|---|---|---|---|---|---|---|---|---|---|
| | 4 | 34 | 64 | 94 | 124 | 154 | 184 | 214 | 244 |
| 1 | 8,74 | 11,37 | 22,67 | 20,49 | 24,60 | 25,96 | 27,36 | 30,12 | 30,93 |
| 2 | 13,34 | 17,71 | 23,60 | 24,10 | 25,94 | 28,12 | 28,50 | 33,10 | 38,98 |
| 3 | 5,68 | 12,46 | 13,51 | 11,15 | 13,56 | 22,67 | 24,67 | 27,12 | 28,43 |

Tabela 2.: Changes in 5-HMF content in enriched sterilized milk during storage. Samples: 1-Milk with added iron, 2-Milk with added sodium caseinate, 3-Milk with added vitamin A, 4-Milk with all three added components, 5-Control.

| Sample No | Content of 5-HMF $(mmol/m^3)$ | | | | | | |
|---|---|---|---|---|---|---|---|
| | Storage period (days) | | | | | | |
| | 5 | 20 | 35 | 50 | 65 | 80 | |
| 1 | 5,06 | 4,81 | 4,73 | 4,81 | 6,56 | 5,69 | |
| 2 | 5,34 | 6,56 | 6,61 | 9,19 | 9,38 | 10,50 | |
| 3 | 2,59 | 2,80 | 3,41 | 1,75 | 4,81 | 4,75 | |
| 4 | 15,75 | 13,30 | 18,38 | 18,31 | 14,79 | 17,50 | |
| 5 | 1,75 | 2,71 | 2,73 | 4,56 | 4,81 | 4,38 | |

Sterilized milks. Content of 5-HMF in enriched sterilized milks is shown in Table 2. Five days after production the lowest content of 5-HMF was in control sample and the highest in milk with all three added components. This difference was significant (1.75 and 15.75 mmol/m$^3$, respectively). Milk with added vitamin A (sample no 3) had a little bit higher 5-HMF content than control, only 2.59 mmol/m$^3$. This level was still higher in milk enriched with iron, and three times higher in milk with added caseinate.

As a rule, in all samples 5-HMF content has increased and after 80 days of storage it was almost doubled. There was one exception in the case of sample no 4. More or less this content was on the same level. The highest increase in 5-HMF content was observed in control sample, 2.5o times; in all enriched milks increase was slightly lower than 2 times ( e.g. milk with added iron 3.06 and 5.69 mmol/m$^3$ after 80 days of storage, etc.).

Content of 5-HMF in milk with added iron significantly increased after 20 and after 65 days of storage, more than 60 %. The same was with control sample, but for milk enriched with sodium-caseinate (sample no 2) this increment was achieved after 50 days of storage.

DISCUSSIONS

The content of 5-HMF in dairy products depends on several factors; among them the most important is heat treatment and concentration of reactants. This was confirmed comparing the content of 5-HMF in control sterilized milk and control milk powder. It was obvious that production of milk powder means more severe heat treatment what resulted, among other changes, in higher 5-HMF content. As pointed out by Patton[1], the presence of iron intensifies Maillard reaction, what is confirmed in all samples where iron is present. This catabolic role of iron in Maillard reactions was obvious especially in sterilized milk enriched with all three components where iron promoted creation of 5-HMF synergistically with caseinate.

Significantly higher content of 5-HMF in sterilized milk with added sodium-caseinate was also expected, with regard that proteins have essential role in this type of reaction. As it is known higher concentration of reactants has consequence in higher yield of reaction products. This also has been proved by the sample no 4.

Adition of vitamin A slightly increased the content of 5-HMF in sterilized milk, but significantly in milk powder what was really surprising.

On the basis of the results for 5-HMF content during storage, it can be undoubtedly concluded that storage at room temperature promotes creation of components appearing in Maillard reactions. In sterilized milk it was logically expected, as the water content was enough for carrying out the Maillard reactions. In the case of milk powders although humidity was less than 5%, reactions have still run in spite of some literature data[8] that no conditions for reactions were achieved in such low moisture content. This could be caused by the way of samples packaging.

References:

[1] S. Patton : J. Dairy Sci., 1955, 38, 457.

[2] L.G. Turner, H.E.Swaisgood, A.P.Hansen : J. Dairy Sci., 1978, 61, 384.

[3] M. Carić : "Technology of concentrated and dried dairy products", Faculty of Technology, Novi Sad, Chapter 3, p. 38.

[4] B. Milić, M. Carić, B. Vujičić : "Nonenzymatic browning reactions in food products", 1983, in manuscript.

[5] J. De Vilder, R. Martens, M. Naudts : Milchwissenschaft, 1979, 34, 78.

[6] J. De Vilder : Le Lait, 1981, 61, 49.

[7] M. Keeney, R. Bassette : J. Dairy Sci., 1959, 42, 945.

[8] J.W.G. Porter,: "Food quality and nutrition, Proceesings of COST Seminar, Dublin, 1977.

[9] E. Ludvig : Die Nahrung, 1979, 23, 707.

[10] E. Ludvig : Die Nahrung, 1980, 24, 399.

[11] E.G. Samuelsson, P. Nielsen : Milchwissenschaft, 1970, 25, 541.

[12] A.B. Möller, A.T.Andrews, G.C. Cheeseman : <u>J.Dairy Res.</u>,1977, <u>44</u>, 267.

[13] A.T. Andrews : <u>J. Dairy Res.</u>, 1975, <u>42</u>, 89.

[14] R. Samel, W.V. Weaver, D.B. Gammack : <u>J.Dairy Res.</u>, 1971, <u>38</u>,323.

[15] J. Korolczuk, I. Kwasniewska, W. Szkilladz, B. Secomska, A. Swiatek : XX International Dairy Congress, Paris, 1978.

# Fat Samples of Certified Composition for Analytical Calibration

By S. P. Kochhar[1]*, J. B. Rossell[1], and P. J. Wagstaffe[2]

[1]LEATHERHEAD FOOD RESEARCH ASSOCIATION, RANDALLS ROAD, LEATHERHEAD, SURREY, U.K.

[2]COMMUNITY BUREAU OF REFERENCE, E.E.C., BRUSSELS, BELGIUM

The objective of this work for the Community Bureau of Reference (BCR) is to prepare a number of edible oil reference materials which will be eventually certified for a range of analytical properties. Four thousand brown glass printed ampoules, 5 ml size, of each of three fat materials (see photograph) were filled under $N_2$ at the Leatherhead Food RA using an automated filling and sealing machine. Measurements of peroxide value, induction period at $100^{\circ}C$, FFA, etc., were used to select the best batch of each fat type. To ensure good stability on storage, TBHQ (250 mg/kg fat) was added to each material prior to filling. BHT and BHA (each 100 mg/kg fat) and vanillin, $\beta$-sitosterol and stigmasterol were also added to the anhydrous milk fat in accordance with the EEC Regulation No 262/79. Fatty acid compositions of the parent oils and their 50:50 blends were determined by GLC.

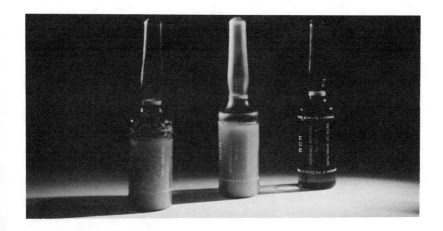

Ampoules of anhydrous milk fat, a beef/pig fat blend (50:50) and a soya/maize oil blend (50:50) prepared for the BCR

The refractive index was determined on 40 evenly spaced and 40 randomly selected ampoules (plus a few corresponding to temporary stoppage of the filling machine) from each of the three batches. The specific extinction values (conjugated dienes at 232 nm and trienes at 268 nm) on half of these samples were also determined. These results were used to confirm homogeneity of the batches of ampouled materials.

Statistical analyses showed that there was no significant difference in the contents of each set of ampoules. All the data were found to fall within the 99 % confidence limits.

An inter-laboratory study on a number of analytical properties is near completion, and storage stability testing of the ampouled materials is now underway.

# The Seasonal Fluctuation of Milk Constituents in Austria (Lactose, β-Carotine, Semimicro Butyric Acid Value, Butyric Acid, *trans*-Fatty Acids, and Total Fat Content)

By W. Pfannhauser

RESEARCH INSTITUTE FOR THE FOOD INDUSTRY, A-1190 WIEN, BLAASSTRASSE 29, AUSTRIA

The starting point of our investigation was the experience that due to fluctuations in milk fat content the addition of dry whole milk powder to e.g. chocolate products has to be increased in slight excess to meet the requirement of legislation.

During a period of 15 months a total of 503 dairy samples from eight different areas of Austria were investigated.

A statistical evaluation was made to find out whether some milk constituents vary during the seasons.

Lactose varies from about 5,3 % in week 50 (mid of December) to 4,7 % in week 35 (end of August).

The content of trans-fatty acids ("trans-index") formed by microbial metabolisation in rumen was in winter lower than in summer. An exception was found only in one dairy possibly due to feeding influenced by the weather conditions.

Butyric acid content lies between 3,5 and 4 % on total milk fat basis and does not vary significantly. There has been found no correlation between "Semimicro Butyric Acid Value" (OICC Methode Nr. 8a) and butyric acid content.

ß-Carotine content in milk is influenced by carotine content of feed. There has been observed a maximum within week 41 (beginning of October) and a minimum at week 4 (end of January) but there were big differences between different dairies. Medium values are 5,38 to 6,30 mg ß-carotine/kg milk fat.

The influences of race, feeding, habits, climate and age of the cows will be discussed.

# Capillary Gas Chromatography, a Valuable Improvement of GC for the Analysis of Dairy Products

By M. Carl

INSTITUTE FOR MILK RESEARCH, HIRNBEINSTRASSE IO, D-8960 KEMPTEN, WEST GERMANY

In the last 3 years fused silica capillary columns with cross-linked nonextractable stationary phases became commercially available. Gas chromatographic equipment with advanced pneumatic and oven designs and powerful electronic integrators are also available at reasonable prices.

The superior quality of these columns not only in terms of separation efficiency, but also in terms of intertness makes them a powerful tool in routine gas chromatography of dairy products, especially for the following practical reasons:

- More information is available in one chromatogram with the same amount of work (e.g. for sample preparation etc.); in many cases a "preseparation" which would be necessary for packed column gc can be avoided.

- Compounds to be determined can be better distinguished from interferences, which especially helps to secure correct identification in trace analysis.

- It is not necessary to have tailor-made columns for almost each problem and therefore not necessary often to change (and again optimize) columns. This considerably saves time which can be spent for a careful optimization and control of the whole gas chromatographic system.

- It is easily possible to install parallel columns in the same injection port (e.g. for identification or confirmation purposes).

Except for the analysis of permanent gases capillary gc completely replaced packed column gc in our laboratory for the above reasons. Practical experience of several years has shown that with the injection port (split/splitless)-capillary column (non-polar, slightly polar, polar)-detector (FID, ECD, TEA, MS) configurations OC-pesticides, PCB's, free fatty acids, fatty acid methyl esters, ethanol and other solvents, sterols, sugars, benzoic and sorbic acids, nitrosamines etc. can be economically analyzed with a high degree of both qualitative and quantitative reliability.

# Determination of Orotic Acid and of Other Carboxylic Acids in Milk and Dairy Products by HPLC

By P. Lavanchy and G. Steiger*
FEDERAL DAIRY RESEARCH INSTITUTE, CH-3097 LIEBEFELD-BERN, SWITZERLAND

Introduction :

Organic acids supply important information on the process of metabolism of milk products and on their quality.

Preparation of sample :

5 g milk or cheese are homogenized with 5 ml water and 20 ml acetonitrile. After centrifugation (10 min, 19600 N) an aliquot is filtrated through a teflon filter (0.45 µm) and then injected into HPLC.

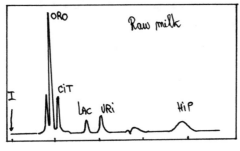

Fig. 1

HPLC conditions :
Packing : Aminex HPX-87 (Bio-Rad Richmond, CA)
Column size : 300 x 7.8 mm
Column temperature : 50°C
Eluent : 10 ml acetonitrile + 990 ml $H_2SO_4$ 0.009 N
Flow rate : 0.6 ml $min^{-1}$
Injection volume : 25 µl
Detector : UV 220 nm, range 0.016 Aufs

Characteristic results (mg/kg)

|  | ORO | CIT | LAC | URI | ACE | PRO | BUT | HIP |
|---|---|---|---|---|---|---|---|---|
| Raw milk | 73.3 | 938 | 91.5 | 16.5 | – | – | – | 36.6 |
| Yoghurt | 75.5 | 725 | 16280 | 30.7 | 110 | – | – | – |
| Kefir | 89.2 | 319 | 20262 | 35.8 | 957 | – | – | 6.2 |
| Sour cream | 45.4 | 130 | 9215 | 15.4 | 880 | 175 | – | – |
| Swiss Sapsago cheese | 15.2 | 10.4 | 2520 | 2.2 | 1420 | 2530 | 14128 | 20.8 |

Repeatability (ORO in raw milk) : 5.6
Recovery : 90-98 %

Literature :

MARSILI, R.T., J. Chromatogr. Sci., 19 (9), 491 (1981)
COUNOTTE, G.H.M., J. Chromatogr., BioMed, 276, 423 (1983)

# The Estimation of Casein in Milk by an Enzyme Linked Immunosorbent Assay

By J. H. Rittenburg, A. Ghaffar, C. J. Smith, S. Adams, and J. C. Allen
RESEARCH DIVISION, THE NORTH EAST WALES INSTITUTE, DEESIDE, CLWYD CH5 4BR, U.K.

A study was undertaken to examine the potential use of an enzyme linked immunosorbent assay (ELISA) to quantify the amount of casein in bovine milk. Current methods of casein analysis are time consuming, complex and at best only give an indirect indication of casein content. By employing antibodies specific to the casein proteins it becomes possible to measure casein directly and to reduce the current assay time of hours or even days down to minutes. The analytical advantages of the ELISA approach has been well proven in the clinical diagnostic field. A number of these assays, now in common use, satisfy the need for a rapid, inexpensive and highly specific assay which is easy to perform.

In the present study, a preparation of whole casein was used to immunize a group of rabbits. Antisera specific for the casein proteins was collected and the casein specific immunoglobulin was isolated. The purified antibody was then adsorbed onto the surface of plastic wells of a microtiter plate. Peroxidase was covatently coupled to casein to give the other essential element of the ELISA. Using these reagents and several standard solutions of known casein concentration, a competitive ELISA could be performed in less than 30 minutes.

In the first step of the assay the milk sample is serially diluted to give a casein concentration which falls within the range of the standard casein solutions. In our current test this means a 1:5000 milk dilution.

The next step is to add a known quantity of the casein–peroxidase conjugate to the diluted milk sample and also to the standard casein solutions. Each of these mixtures is then added to the antibody coated wells of the microtiter plate. After a short incubation period (10 min), the solutions are washed out of the wells and a chromogen solution is then added. Following another short incubation period the optical densities of the wells are measured and a standard response curve is derived from the known casein values. Using the standard curve the casein concentration in the milk sample can be estimated. It was observed that at casein concentrations above 20 $\mu$g/ml there was over 90% inhibition of colour, whilst the lower limit of sensitivity occurred at approximately 0.5$\mu$g/ml. Tests in which whey proteins were included demonstrated the specificity of the assay to casein as no cross reactions were observed.

The present study demonstrates the potential application of an ELISA for measuring casein in milk. By having the means for rapidly estimating casein levels it may then be possible to screen milk prior to processing thus allowing the various processing parameters to be optimized for milks of varying composition.

# Gas Chromatographic Method for the Determination of Solanidine in Bovine Milk from Lactating Cows Fed Potato Waste

By R. J. Bushway, D. F. McCann, J. L. Bureau, and A. A. Bushway

DEPARTMENT OF FOOD SCIENCE, UNIVERSITY OF MAINE, ORONO, MAINE 04469, U.S.A.

Today processed potatoes represent 57% of the total potato production in the United States and the percentage continues to rise. Considerable waste is generated during processing and one means of utilizing it is the production of animal feed like tater meal which is a potato waste product comprised of cull potatoes and process by-products such as peel, screen, drum, French fry and filter cake wastes. These waste products contain high levels of naturally occurring toxicants, potato glycoalkaloids, which at elevated levels can impart a bitter flavor to tubers and are toxic to man and animals. Much of the potato waste is fed to dairy cattle as a feed supplement. It has been shown that rumen microorganisms cleave glycoalkaloids yielding solanidine. Although very little research has been performed on the toxicological and physiological properties of solanidine, there is evidence that suggests that it is similar to the glycoalkaloids as far as bitterness and possibly toxicity. Because of the extensive use of potato waste as animal feed for dairy cows, the possible detrimental effects of solanidine and the large consumption of milk and milk products, an investigation was conducted to develop a gas-chromatographic method to quantify solanidine in whole milk and to determine if a feed ration containing potato by-products resulted in the passage of solanidine into the milk of lactating dairy cows. Milk, freeze dried, was saponified followed by partition into toluene. Quantification was achieved with a nitrogen-phosphorous detector. Spiking studies indicated approximately 93% recovery of added solanidine at levels of 1.12, 0.56 and 0.28 ppm with a detection limit of 0.14 ppm. A study was conducted to determine if solanidine was present in milk from lactating cows consuming diets containing 10 and 20% tater meal (A potato waste product containing 40 to 60 ppm solanidine in the form of glycoalkaloids). No detectable amount of solanidine was observed in any of the milk samples. Results were confirmed by thin-layer chromatography.

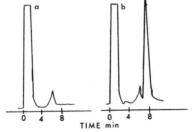

Figure 1. Gas Chromatograms of (a) 20% tater meal milk sample and (b) milk spiked with 1.12 ppm solanidine.

# Determination of Free Fat in Cream for the Assessment of Damage Caused to Fat Globules by Processing

By A. Fink and H. G.Kessler

MUNICH INSTITUTE OF TECHNOLOGY, FREISING-WEIHENSTEPHAN, D-8050 FREISING, WEST GERMANY

Cream (determination of free fat)
The purpose of the determination of free fat in cream (e.g. 30% fat) is to assess the demage caused to fat globules by processing (e.g. mechanical and thermal treatments). A. modification of the centrifugation method is described in which the free (i.e. membrane-less) fat is separated by centrifugation in butyrometers.The variables that were examined and standardized were: the dilution medium, the degree of dilution, the pH value of the diluent and the centrifugation time.

Another method investigated is the determination of free fat by the extraction methqd which, additionally, measures the fat contained in globules whose membranes are porous or damaged. This method was also modified to make it suitable for cream and the variables standardized were: the nature of the extraction medium, the extraction time and the type of the extraction vessel.

The significance of the results of both methods was investigated in a number of experiments. The centrifugation method can immediately detect mechanical damage to the fat globules, e.g. caused by pumping, while the extraction method is yet unable to detect the relatively small amounts of free fat present at that stage.The extraction method detects damage to the fat globule membrane, produced, for instance, by homogenization or heating immediately after processing, while the centrifugation method is as yet insensitive. Centrifugation is a suitable method only after a prolonged period of storage when enough free fat has escaped from the interior of the fat globules. The extraction method gives, therefore, an early indication of the sort of damage that will cause undesirable creaming on prolonged storage of cream.

# Determination of Heavy Metal Contaminants in Milk and Milk Products

By M. Carl

INSTITUTE FOR MILK RESEARCH, HIRNBEINSTRASSE 10, D-8960 KEMPTEN, WEST GERMANY

Iron, copper, lead, tin, cadmium and mercury have been widely discussed as the most interesting heavy metal contaminants in milk and milk products, iron and copper for "technical" reasons, the others for health reasons.

Up to now reliable information on the contents of these metals in milk and milk products is rare, the main reasons for that being not the methods and techniques itself, but their performance in different laboratories.

For the purpose of the dairy industry spectrophotometric, atomic absorption (flame, graphite oven and cold vapor) and differential pulse anodic stripping voltammetric techniques can be recommended. IDF and ISO standards for the determination of copper and iron based on spectrophotometry have been published after thorough collaborative work and have been recently (in the BCR certification exercise) proven (in experienced laboratories) to be very reliable.

Most important steps of the analytical procedures used in heavy metal trace analysis are calibration and contamination control. Contamination can be sufficiently reduced by

- cleaning acids by isothermal distillation in quartz,

- cleaning other reagents by extraction or extraction chromatography,

- using vessels of quartz, polytetrafluoroethylene or polyethylene and cleaning them by steam of (diluted) acids,

- using high purity water,

- working in closed systems, as far as possible, and protecting open systems by clean laminar air flows.

Quality control in the laboratory is very essential. For that purpose 3 milk powders contaminated at different levels will be available as reference materials in the near future from the Community Bureau of Reference, Brussels. They have been successfully analyzed by competent laboratories using the best available techniques and will be certified for their contents of a.o. iron, copper, lead, cadmium and mercury.

# IDF Disc Assay (*Bacillus stearothermophilus*) and Charm Test[R] Comparisons with Low Concentrations of Beta-Lactams (0.0025—0.006 I.U./ml) and Mycins

By S. E. Charm

DEPARTMENT OF CHEMICAL ENGINEERING, TUFTS UNIVERSITY, MEDFORD, MA. 02155, U.S.A. and PENICILLIN ASSAYS INC., 36 FRANKLIN STREET, MALDEN, MA. 02148, U.S.A.

The IDF Disc Assay is reported to detect concentrations of beta-lactam drugs as low as 0.0025 I.U/ml penicillin. A number of countries are adopting an action level equivalent to 0.005 I.U/ml or less (e.g. Japan, New Zealand, Australia, Canada).

In the U.S. a slightly modified IDF test is employed where the action level is a 16 mm inhibition zone, equivalent to about 0.006 I.U/ml.

B. stearothermophilus is employed in both CHARM and IDF tests. In the CHARM TEST it acts as a specific binder for antibiotics, permitting a competitive assay when a tagged antibiotic is added. In the IDF test, growth is inhibited by the antibiotic.

There is a direct correlation between binding beta-lactam drugs and growth inhibition of various microorganisms.

Mycins bind to sites on cell ribosomes permitting their detection with the CHARM TEST.

Beta-lactam and mycin drugs can be detected simultaneously in milk within 9 minutes using the CHARM TEST and a scintillation counter.

With standard CHARM TEST equipment, it requires 20 minutes to measure 0.0025 I.U/ml, while concentrations greater than 0.005 I.U/ml can be determined in less than 15 minutes. The IDF test requires 2.5 to 3 hours.

In the U.S. more than 70% of milk is screened with the CHARM TEST before tankers unload at the processing plant  (May '83 <u>Dairy Record</u>).

# Determination of Retinol (Vitamin A) in Dry-mixing Fortified Dried Skim Milk with HPLC: Sources of Variation

By R. Van Renterghem and J. De Vilder
GOVERNMENT STATION FOR RESEARCH IN DAIRYING, MELLE, BELGIUM

Rather high variations are found when determining the retinol content of dried skim milk (DSM), fortified by the dry-mixing-procedure, using a HPLC method. The sources of these variations were investigated.

The analytical method used is a modification of the IDF project E 46 - doc 11 - october 1981.

Twenty g DSM are saponified with 80 ml ethanol, 20 ml KOH solution (50%) and 20 ml sodium ascorbate solution (20%). Anthracene is added as an internal standard (2 ml of a 100 mg per 100 ml solution). The saponifaction is performed on a steam bath during 45 minutes. Preliminary tests showed this quantity of sample to be suitable for the given quantity of saponifying liquid; samples exceeding 30 g regularly enhanced clump-formation. A saponification time between 30 and 60 minutes gave the best results.

A saponification time of 45 minutes was adopted. After saponification, extraction was carried out once with 60 ml water + 20 ml ethanol + 80 ml diethyl ether and twice with 20 ml ethanol + 40 ml pentane.

The combined ether-pentane extracts were washed with 3 x 40 ml aqueous-alcoholic KOH soln (3 g KOH in 10 ml ethanol made up to 100 ml with water) and with 40 ml portions of water until neutral to phenolphthalein. The extract was dried over filter paper cuttings and made up to 250 ml with pentane. Fifty ml of this solution were evaporated to dryness (rotary evaporator, max. 40°C) and the residue was solubilized in 5 ml of HPLC-methanol. 20 µl were injected in the HPLC, equipped with a 25 cm x 4.6 mm column, packed with 10µC8.

As mobile phase a mixture of methanol-water: 90-10 was used; the flow was 2 ml/min and detection was at 325 nm. Total variation of the analytical results is considered to be the compilation of

1. sampling variation (due to inhomogeneity)
2. sample treatment variation (saponification, extraction, etc.)
3. HPLC variation (mainly due to variations of the injection on a variable sampling loop).

Coefficients of variation (C.V.) are used here as a measure of variation. The HPLC variation was determined by injecting 5 retinol solutions 10 times each. The mean C.V. was 1.25 %.

The variation of chromatography + sample treatment was estimated by determining the retinol content of seven samples of wet-fortified DSM. The use of wet-fortified DSM samples was supposed to eliminate most of the variation due to the inhomogeneity of the samples as encountered in dry-mixing fortified powders. Each sample was analysed five times. The mean C.V. was 2.00 %. The overall C.V. of the analysis as determined by analysing seven samples of dry-mixing fortified DSM (analysed five times each) amounted to 8.32 %.

From these data it is concluded that the main source of variation when determining retinol in dry-mixing-fortified DSM, using HPLC, is the inhomogeneity of the sample.

Recently, as stated in IDF doc 165-1983, a working group studying this analytical method drew conclusions in agreement with our results and proposed to dissolve the sample in hot water before saponification, in order to overcome problems related to the inhomogeneity of the sample.

# Determination of Furosine as an Indicator of Early Maillard Products in Heated Milk

H. F. Erbersdobler[1], H. Reuter[2], and E. Trautwein
[1]UNIVERSITY OF KIEL, DÜSTERNBROOKER WEG 17-19, D-2300 KIEL I, WEST GERMANY
[2]FEDERAL STATION FOR MILK RESEARCH, HERMANN-WEIGMANN STRASSE I, D-2300 KIEL I. WEST GERMANY

The heating of proteins in the presence of glucose or lactose results in the formation of fructoselysine or lactuloselysine (Galactose-Fructoselysine), in which the sugars are linked at the ε-amino group of lysine. Both compounds, which are very unstable to acid hydrolysis, can be estimated by analysing furosine, which is formed during the hydrolysis with strong hydrochloric acid. We detected this useful indicator many years ago (Erbersdobler and Zucker 1966) and could meanwhile show that the fructoselysine moiety occurs in considerable amounts in many heat damaged foods.

We have previously analysed furosine on a short column (3o-5o mm x 4-9 mm) but offered some years ago also a simultaneous determination of lysine, lysinoalanine and furosine by using a long (16o mm x 4 mm) column and a 4 buffer and 3 temperature programme. By modifying this procedure it was possible to obtain a threefold increase in the sensitivity of the method, which can be forced up by using a fluorimetric method. Also the specificity could be improved and furosine could be separated from an unknown peak, which appeared in the milk and seems to be of endogenous origin. The figure 1 shows some preliminary results of our analyses in UHT-milk samples.

Analytical procedure
- Buffers - temp.time

| pH | Na⁺M | | |
|---|---|---|---|
| 2.95 | 0.16 | 53°C | 3′ |
| 4.1O | 0.22 | 53°C | 5′ |
| 4.1O | 0.22 | 59°C | 2o′ |
| 6.4O | 0.46 | 59°C | 1o′ |
| 6.4O | 1.16 | 59°C | 3o′ |
| NaOH | 0.4O | 69°C | 15′ |
| 2.95 | 0.16 | 53°C | 35′ |

Buffer flux:2o ml/h
Ninhydrin flux:1oml/h

Resin: MCI Gel, 1OF, Mitsubishi, Japan

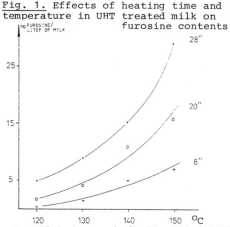

Fig. 1. Effects of heating time and temperature in UHT treated milk on furosine contents

Erbersdobler, H.F. and Zucker, H. Milchwissenschaft 21, 564 (1966)

Erbersdobler, H.F., Holstein, B. and Lainer, E. Z.Lebensm.Unters. Forsch. 168, 6 (1979)

# Fractionation of Radioactivity in the Milk of Goats Administered [14]C-Aflatoxin B1

By T. Goto[1,2] and D. P. H. Hsieh[1]

[1]DEPARTMENT OF ENVIRONMENTAL TOXICOLOGY, UNIVERSITY OF CALIFORNIA, DAVIS, CA. 95616, U.S.A.

[2]NATIONAL FOOD RESEARCH INSTITUTE, TSUKUBA, JAPAN

A detailed fractionation of radioactivity in the milk of goats administered 14C-aflatoxin Bl (AFB) was performed. Each goat was administered 100 or 150 µCi of AFB (146 or 209 µCi/µmole) i.v. or orally. Milk samples were collected at time intervals and were frozen (-80°C) until analysis. Each milk sample was fractionated into four fractions: ether extractable, proteins (70% ethanol precipitation), dichloromethane extractable, and water-alcohol soluble. The milk collected in the first 24 hr contained radioactivity equivalent to 0.45-1.1% of the initial dose, of which 47-85% was secreted in the first 6 hr. The milk collected in the next four days contained 0.02-0.28% of the initial dose. AFB and aflatoxin Ml (AFM) were analyzed by TLC and quantitated by fluorodensitometry. TLC plates were developed with toluene-ethyl acetate-90% formic acid (6:3:1 v/v/v). Radioactivity was measured by liquid scintillation counting. As shown in Table I, most of the radioactivity was detected in the dichloromethane extractable fraction, most of which was attributable to AFM as shown by direct scintillation counting of AFM isolated by TLC. No AFB was detectable either by fluorodensitometry or by measurement of radioactivity. The bacterial mutagenicity of various fractions was tested using a modified Ames Salmonella/microsome mutagenicity assay with increased sensitivity. No fraction contained mutagenicity greater than that of AFM on an equivalent radioactivity basis. Our results indicate that the major metabolite of AFB in the milk of goats administered this toxin is dichloromethane extractable AFM. Other metabolites, including conjugates, are of minor significance.

Table I. Distribution of Radioactivity in the Fractions of Goats Milk Collected in 6 hr Following 14C-aflatoxin $B_1$ Dosing

| Goat | Radioactivity (dpm/g milk) | | | |
|------|-------|---------|-----------------|--------------|
| | Ether | Protein | Dichloromethane | Water-ethanol |
| #1 (i.v.) | 9 | 278 | 2,845 | 217 |
| #2 (i.v.) | 63 | 164 | 4,827 | 308 |
| #3 (oral) | 47 | 28 | 619 | 46 |
| #4 (oral) | 25 | 71 | 1,961 | 42 |

# Trace Metal Levels in Yugoslavia Dairy Products

By L. Kršev

BIOTECHNOLOGICAL FOODS FACULTY, PIEROTTIJEVA 6, 41000 ZAGREB, YUGOSLAVIA

Trace metal conc. were determined in samples of market milk
and a range of dairy products (cream, dried milk, butter,
yoghurt, dairy dessert)from 5Yugoslavia dairy factories, 1
sample of each dairy product being taken from each factory
once every 2 months over a 12 month period. Trace metal conc. were determined by AAS in samples. Mean values for 23
market milk samples were (p.p.m) : Cr, 0,015 $\pm$ 0,0004 ;
Cu, 0,199 $\pm$ 0,0047 ; Fe, 0,637 $\pm$ 0,0184 ; Pb, 0,029 $\pm$ 0,001 ;
and Zn, 5,071 $\pm$ 0,072. Results for Hg were all <0,001 p.p.m.
No significat geographical differences were found, but significant seasonal differences were found for Cr, Fe and
Pb in the winter. Ranges of metal conc. (p.p.m) in the milk
products were : Cr, 0,015 (cream) to 0,952 (yoghurt) ;
Cu, 0,075 (butter) to 0,915 (dried milk) ; Pb, 0,0231 (cream)
to 0,071 (dried milk) ; Zn, 312 (cream) to 15,01 (yoghurt) ;
Hg conc. were < 0,001.
All conc. trace metal were comparable with those reported
in other countries.

# The N-15 Technique Used for Investigation of the Fate of Nitrate in Cheese

By L. Munksgaard
THE DANISH GOVERNMENT RESEARCH INSTITUTE FOR THE DAIRY INDUSTRY, DK-3400
HILLERØD, DENMARK

In order to investigate what becomes of nitrate in cheese
during the ripening period, cheese was produced under addition of nitrogen-15-labelled nitrate. Nitrogen-15 is a stable
(i.e. not radioactive), naturally occuring nitrogen isotope,
that can be distinguished from nitrogen-14 by mass spectrometry.

During cheese ripening for 16 weeks the nitrate breakdown
was studied by mass-spectrometric determination of the content of N-15 and by determination of nitrate in the cheeses.
Further the cheeses were separated into fat, protein and
non-protein-nitrogen and the N-15 content of these fractions
was determined.

The amount of N-15 evaporating from the cheeses as gases
was studied by gaschromatography and mass spectrometry of
headspace from closed glass-containers in which some of the
cheeses were kept under simulated cheese-store conditions
during the ripening period. As atmospheric nitrogen in the
containers had been replaced by helium it was possible to
detect even small increases in the amount of N-15 in $N_2$ and
$N_2O$ in the headspace.

# Lactulose as an Indicator of the Severity of the Heat Treatment of Longlife Milk

By G. R. Andrews

NATIONAL INSTITUTE FOR RESEARCH IN DAIRYING, SHINFIELD, READING, BERKSHIRE
RG2 9AT, U.K.

Lactulose (4-σ-β-D-galactopyranosyl-D-fructose) was determined enzymatically[1] in commercial pasteurized, UHT and in-container sterilized milk samples and also in pilot plant processed milks to investigate the possibility of using it to distinguish UHT and in-container sterilized milks.

For some commercial samples, the total chemical change caused by the heat process was quantified by calculating, from the time/temperature profile, a dimensionless integral, $C^*$, based on the kinetics of the thermal degradation of thiamine.[2]

Raw milk was processed in an indirectly heated UHT pilot plant to give nine milk samples of $C^*$ from 0.2 to 1.8 in steps of 0.2.

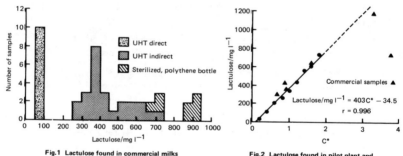

Fig.1 Lactulose found in commercial milks

Fig.2 Lactulose found in pilot plant and commercial milk samples

No lactulose was found in pasteurized milk. The relationship between $C^*$ and lactulose was confirmed by the fit of the commercial samples. The anomalous result in fig.2 was for sterilized milk in a polythene bottle. This is being investigated.

[1] H. Geier and M. Klostermeyer, Zeitschrift für Lebensmittel-Untersuchung und-Forschung, 1980, 171, 443.

[2] H.G. Kessler, 'Food Engineering and Dairy Technology', Verlag A. Kessler, Friesing, Munich, 1981, Chapter 6, p.198.

# Screening for Heavy Metals and Fat Soluble Pesticides in Milk by Simple and Accurate Methods

By S. de Leeuw*, R. Robbiani, and W. Büchi
CPC/EUROPE, QUALITY ASSURANCE CENTRE, LEUTSCHENBACHSTRASSE 46, CH-8050
ZÜRICH, SWITZERLAND

Milk is, especially for children, an important constituent
of the daily food. Its nutritive value is high; the presence
of contaminants, however, like DDT and other fat soluble
pesticides can diminish the value of this nutrient. Heavy
metals like Pb and Cd, which are found almost everywhere in
the environment, can also contaminate milk and its products.
Finally, metals like copper and iron can decrease the shelf-
life of milk. Therefore, the need is given to monitor milk
samples periodically for residues of the above kind. For this
reason, methods for the analysis of heavy metals and residues
of fat soluble pesticides were developed, which have the
merit of being fast  - the results can be obtained before
distribution of the milk, if necessary - , simple, accurate
and non-laborious. The poster shows a description of these
methods, some results, which steps can be automated, and
how quickly the results are obtained. The method for heavy
metals consists of a simple digestion procedure and flameless atomic absorption spectroscopy detection, that for fat
soluble pesticides of a simple clean-up procedure and gas
chromatographic analyis. Contaminant levels well below
current maximum limits can be detected with a good accuracy.

# Determination of Skim Milk Powder Content in Animal Foodstuffs by a Rocket Immunoelectrophoretic Method

By A. Driou, G. Godbillon, and G. Linden*

LABORATORY OF APPLIED BIOCHEMISTRY, UNIVERSITY OF NANCY I,
VANDOEUVRE-LES-NANCY, FRANCE

The purpose of the present work was to investigate a method
for the determination of skimmilk content in foodstuffs; we
have chosen the rocket immunoelectrophoretic method of
Laurell (1966) using a rabbit antiserum to Kappa casein
which has been prepared in our laboratory.  Quantifications
were made by comparing the sample peak heights with a
standard curve obtained from peak heights of known amounts
of skimmilk on the same plate.

The immuno precipitate depended on the antigen; for skimmilk,
two peaks were formed in the rocket; we could see a small
inner peak above the gel hole.  The treatment of skimmilk
by chemical reagents increased, decreased or did not modify
the inner peak height.  A thermic treatment applied on a
skimmilk powder solution, for 45 seconds in a boiling water
bath, gave rise to a single high peak.

The accuracy was better in skimmilk solution (1 p.100)
comparing to the determination of skimmilk content in
animal foodstuffs (5 p.100); the peak was diffused and
difficult to measure accurately.

A rocket immunoelectrophoretic experiment was carried out
with different industrial whey powders.  Immuno precipitates
were formed with a few samples against the antiserum to
Kappa casein; we could see two peaks as in skimmilk.  These
whey powders contained a low quantity of Kappa casein:
contamination by curd, Kappa not coagulated by rennet.

# Quantitative Analysis of Proteinases and Lipase from Psychrotrophic Fluorescent *Pseudomonas* Strains with Enzyme-linked Immunosorbent Assay (ELISA)

By S. E. Birkeland, L. Stepaniak, and T. Sørhaug
DAIRY RESEARCH DEPARTMENT, AGRICULTURAL UNIVERSITY OF NORWAY, ÅS-NLH, NORWAY

Recent time have shown an increased use of immunological techniques for the quantitative anlysis of staphylococcal enterotoxins and other microbial metabolites. Generally there is an expanding interest to exploit the ELISA and other antibody-based assays for specific and very sensitive detection of food components and contaminants.

In dairy science and technology there is a strong demand for sensitive, specific and rapid methods to detect heat resistant, spoilage proteinases and lipases, particularly from Pseudomonas sp. An ELISA procedure was developed for this purpose.

Rabbit antibodies were obtained against purified enzymes of Ps.sp. from milk cultures. The lower detection limits with the ELISA for the hydrolases produced in milk were: 0,25 ng/ml of proteinase $PP_1$ from Ps. fluorescens P1, 2,5 ng/ml of an immunologically unrelated proteinase $PP_{21}$ from Pseudomonas sp. AFT21 and 100 ng/ml of lipase $LP_1$ from Ps. fluorescens P1.

Proteinases serologically identical to proteinase $PP_1$ were produced by most Ps. isolates from raw milk. Some strains showed complex proteinase system comprising enzymes immunologically unrelated to $PP_1$. Immunologically unrelated lipases among or within Ps.strains were not found.

Detectable levels of $PP_1$ in undiluted and diluted milk cultures at $7°C$ appeared when the cell densities reached $10^7$ and $10^6$ CFU/ml respectively. $LP_1$ was detected in both undiluted and diluted milk concurrent with a level of $10^6$ CFU/ml while $PP_{21}$ needed about $10^8$ CFU/ml for detection. The influence of aeration and mixed culture conditions were also studied.

# Indicators of the Heat-treatment Conditions to Which Milk Has Been Subjected

By M. Anderson, E. W. Evans, K. R. Langley, D. J. Manning, G. A. Payne, and E. A. Ridout

NATIONAL INSTITUTE FOR RESEARCH IN DAIRYING, SHINFIELD, READING, BERKSHIRE RG2 9AT, U.K.

Heat treatments have evolved for milk and cream, as for most foods, that enable a safe product with acceptable shelf-life to be produced without undue loss of nutritive value or of sensory acceptability. Methods for retrospectively assessing the heat treatment given to a product could have a number of uses including the monitoring of minimum standards and the identification of multiple, unconventional or reprocessing treatments.

The classical method of monitoring severity of heat treatment by the extent to which micro-organisms or their spores are destroyed requires considerable care, and is not applicable in retrospect to a product "off-the-shelf". This poster describes an investigation into the possibility of assessing whether the duration and temperature of the heat treatment to which a product has been subjected during processing can be determined, at a later date, from its chemical composition.

Tests for effectiveness of pasteurisation and sterilization of milk depend on assays of destruction of native enzymes and of changes in the casein. They give no indication of the specific duration or of the specific temperature of the heat treatment. A satisfactory test must cope with changes in product composition, that occur in most natural products, processing and with storage conditions.

These objectives can be met only by having several parameters that are sensitive to heat over the range of heat treatments of concern, and moreover the rates of reactions by which these parameters are changed must differ significantly.

The NIRD, sponsored by MAFF, carried out an experiment jointly with the Milk Marketing Board of England and Wales, to assess the application of such an approach to the diagnosis of heat treatment of milk. Milk was heated in an experimental plant over a range of temperatures between 75$^\circ$C and 150$^\circ$C for durations between 2 and 300 secs. Several parameters were measured on these heated milks, including viscosity, casein micelle size, nitrogen distribution, formation of volatiles, e.g. heptanone, rennet clotting time (RCT) denaturation of the whey proteins, formation of lactulose, and heat number. Some of these parameters have a common basis in chemistry, e.g. heat number, nitrogen distribution and whey protein denaturation, but did not in practice have the same reaction kinetics.

Micelle size distribution was determined by controlled pore glass chromatography, heptanone by GLC and whey proteins and lactulose by HPLC.

Mathematical models of whey protein denaturation, heat number, lactulose production and rennet clotting time accounted for a high proportion ($\sim$85%) of the variances, and most of the relationships between ln (time) and 1/Temperature were linear.  These models can be used to solve for duration and temperature of heat treatments over the range between pasteurisation and sterilization treatments.

Further work is needed to assess the extent to which the approach is confounded by variation in the fat content of the milk and by processing and storage conditions.